BOATS OF SOUTH A!

Based on recent fieldwork in India, Bangladesh and Sri Lanka, this book documents some of the vast array of traditional boats used in the sub-continent today for fishing and other coastal and riverine tasks. Written in non-technical language, it sets new standards for the documentation of water transport, and introduces styles of boat-building which are unlikely to be found outside the sub-Continent.

An important book for ethnographers, maritime archaeologists and historians, *Boats of South Asia* sets the boats in their geographical environment and their technological and social context.

Seán McGrail was in the Royal Navy for 22 years. He was Chief Archaeologist at the National Maritime Museum, Greenwich (1976–86) and Professor of Maritime Archaeology at the University of Oxford (1986–93). He is currently Visiting Professor in the Department of Archaeology, University of Southampton. He has written several books, including *Ancient Boats of Northwest Europe* (1998) and *Boats of the World: From the Stone Age to Medieval Times* (2001).

Lucy Blue is Lecturer in Maritime Archaeology at the University of Southampton. **Eric Kentley** is a maritime ethnographer, formerly at the National Maritime Museum, Greenwich. **Colin Palmer** is a naval architect, specialising in boats. **Basil Greenhill** was Director of the National Maritime Museum, Greenwich, from 1967 to 1983.

ROUTLEDGECURZON STUDIES IN SOUTH ASIA
Editor Michael Willis
The British Museum

RoutledgeCurzon publishes a monograph series in association with *The Society for South Asian Studies*, London.

BOATS OF SOUTH ASIA
Seán McGrail

MUSLIM ARCHITECTURE OF SOUTH INDIA
Mehrdad Shokoohy

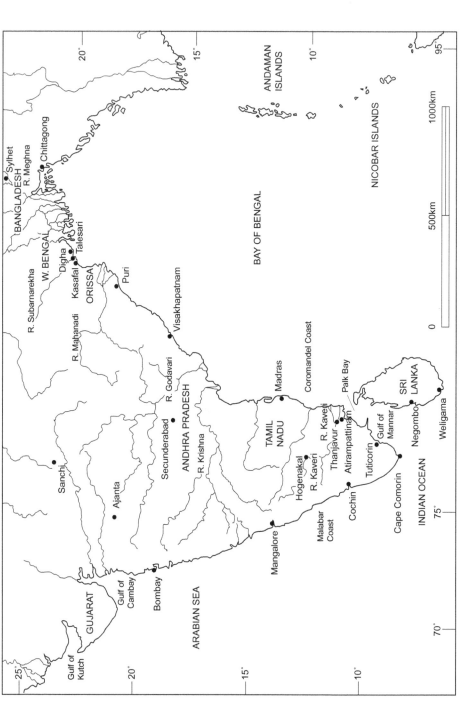

Map of South Asia showing the location of key sites in Sri Lanka, Tamil Nadu, Andhra Pradesh, Orissa, West Bengal and Bangladesh

BOATS OF SOUTH ASIA

Seán McGrail

with
Lucy Blue, Eric Kentley and
Colin Palmer

Introduction by Dr Basil Greenhill

Routledge
Taylor & Francis Group

LONDON AND NEW YORK

First published in 2003 by Routledge

2 Park Square, Milton Park, Abingdon, Oxon OX14 4RN
711 Third Avenue, New York, NY 10017, USA

Routledge is an imprint of the Taylor & Francis Group, an informa business

First issued in paperback 2016

Typeset in Garamond by LaserScript Ltd, Mitcham, Surrey

British Library Cataloguing in Publication Data
A catalogue record of this book is available from the British Library

Library of Congress Cataloging in Publication Data
McGrail, Seán
Boats of South Asia / Seán McGrail with Lucy Blue ... [et al.].
p. cm. – (RoutledgeCurzon studies in South Asia)
Includes bibliographical references and index.
1. Maritime anthropology–South Asia. 2. Boats and boating–South Asia. 3.
Boatbuilding–South Asia. 4. South Asia–Social life and customs. I. Title. II.
RoutledgeCurzon-IIAS Asian studies series.
GN635.S57 M34 2002
623.8′2′00954–dc21   2002031604

ISBN 978-0-415-29746-2 (hbk)
ISBN 978-1-138-96438-9 (pbk)

CONTENTS

CONTENTS

ILLUSTRATIONS

The copyright of illustrations is held by the authors except where stated otherwise in captions.

TABLES

CONTRIBUTORS

Dr Lucy Blue, BA DPhil, Centre for Maritime Archaeology, University of Southampton.

Dr Basil Greenhill, BA PhD CB CMG FSA FR Hist S., Centre for Maritime Historical Studies, University of Exeter.

Dr Eric Kentley, BA PhD, Consultant, London.

Professor Seán McGrail, MA PhD DSc FSA MIFA, Master Mariner, Centre for Maritime Archaeology, University of Southampton.

Colin Palmer, CEng MRINA, Water Transport Consultant, Bristol.

ACKNOWLEDGEMENTS

The continued support of the Society for South Asian Studies is gratefully acknowledged. We are especially indebted to Dr David MacDowall, Chairman of the Society for most of the period of our research grant, who suggested that we should write this monograph. We are also most grateful to Dr Bridget Allchin, former Editor of *South Asian Studies*, for her guidance, and to Sir Oliver Forster, former Treasurer of the Society, for his advice.

We are also grateful for travel grants from the Department of Archaeology, University of Southampton, throughout the five years, and from the National Maritime Museum during the earlier years.

Dr Basil Greenhill, Director of the National Maritime Museum from 1967 to 1983, pioneered fieldwork in Bangladesh in the 1950s. We are particularly grateful to him for discussions over many years on all aspects of the research described in this book and for his recent invaluable advice on Bangladesh boats, reverse-clinker planking and hulc planking patterns. We have benefited also from discussions on aspects of our research with Dr John Coates, Dr Arne-Emil Christensen, Dr Brad Loeven and Richard Barker.

Dr Brian Durrans and Dr Richard Blurton facilitated the study of Indian hide boats in the British Museum's Department of Ethnography. Information on a Bangladesh model of a *khelma* reverse-clinker boat was kindly made available by Peter FitzGerald of the Science Museum and Dr Graham Parlett of the Victoria and Albert Museum.

Dr Lotika Varadarajan of the National Institute of Science, Technology and Development Studies (NISTADS) first kindled enthusiasm for work in the subcontinent when she visited the Institute of Archaeology, Oxford, in 1993. Dr Himanshu Ray of the Centre for Historical Studies, Jahawaral Nehru University (JNU), New Delhi, introduced us to NISTADS and to the London-based Society for South Asian Studies, and considerably facilitated our first season of fieldwork in Orissa. Umakant Mishra, a post-graduate student at JNU, was our guide and interpreter during that fieldwork and made a significant contribution to the resulting publication (*South Asian Studies* 13: 189–207). Professor Victor Rajamanickam, Head of the Department of Earth Sciences at Tamil University, Thanjavur, who documented some of the rafts and boats of southern India in the 1980s, has been a strong supporter of the Boats of South Asia project since its inception. Our three seasons of work in Tamil Nadu were undertaken under the auspices of his department.

We acknowledge with gratitude the great help received from numerous other individuals in South Asia, especially the following:

Bangladesh
Professor Shahidal Islam	University of Chittagong
Abrar Hossain	University of Rajshahi
David Walker	British High Commissioner, Dhaka

West Bengal
Dr Gautam Sengupta	Director of Archaeology, West Bengal
Prakash Chandra Maity	Directorate of Archaeology, West Bengal

Orissa
Professor Das, Vice-Chancellor	Utkal University, Bhubaneswar
Professor K.S. Behera	Dept of History, Utkal University
Dr L.N. Raut	Dept of History, Berhampur University

Tamil Nadu
Dr N.J. Chandrasakar	V.O.C. College, Tuticorin
A/Professor N. Athiyaman	Centre for Underwater Archaeology, Tamil University
Jasper Uttley	British Council, Chennai
Dr Jean Deloche	École Française d'Extrême-Orient, Pondicherry
Menaka Rodriguez	Christian College, Chennai
Kalai Selvam	Driver and helpful companion for three seasons

Finally, we are most grateful for the co-operation given to us by numerous fishermen and boatbuilders in Bangladesh, West Bengal, Orissa and Tamil Nadu.

Centre for Maritime Archaeology, University of Southampton
14 November 2001

INTRODUCTION

Basil Greenhill

Joseph Conrad's first encounter with the East was dramatic. As he tells it in his semi-autobiographical story 'Youth', he went ashore in one of the ship's boats from the burned-out and abandoned barque *Palestine* – he calls her *Judea* – at Muntok in Sumatra, as it was then called (now part of Indonesia), in 1883. In 'Youth'[1] he writes:

> I pulled back, made fast again to the jetty, and then went to sleep at last ... I opened my eyes and lay without moving. And then I saw the men of the east – they were looking at me. The whole length of the jetty was full of people. I saw brown, bronze, yellow faces, the black eyes, the glitter, the colour of an eastern crowd. And all these beings stared without a murmur, without a sign, without a movement.[2]

My first encounter with the East was somewhat different. In 1950, having travelled from Britain by the old way – on the P&O *Strathmore* through the Suez Canal to Bombay, then by train across the subcontinent to Calcutta – my first real sight of the further East was from a Douglas DC3, a wartime surplus Dakota, as we approached the single runway at what was then Dacca Airport in what is now Bangladesh. And the feature of the scene which fascinated me at once as we came in low over the monsoon-inundated countryside was the boats, scores and scores of them, under squaresails and spritsails, being rowed, sculled and paddled, or some of them towed from the mast head by two or three men on a river bank. I had had no forewarning that we were approaching what was then probably the richest boat culture in the world at the height of its richness.

In four years, on and off, of living in Bangladesh my wife and I met Conrad's silent, friendly, watching crowds many times, indeed whenever we examined and photographed a boat in detail. The study of the boats, 'country boats', as they were known to English speakers, and the way of life around them, the way they were manned, managed, financed, sailed, built and navigated, became our principal leisure interest – and it complemented very nicely the professional reasons for being in Bangladesh. The term 'country

1

boat' incidentally used to have a wider meaning. It applied to any vessel owned and managed by the natives of the subcontinent of South Asia, were she a dinghy or a wooden barque trading in the Indian Ocean or beyond. It didn't describe a type of boat, but her ownership.[3]

Our work in Bangladesh in the 1950s enabled us to record a number of aspects of the water transport of that country which had not been brought to wider attention before. We were able to describe the remarkable survival of expanded logboats, some up to forty or fifty feet in length, as sailing cargo carriers in the eastern Delta area and on its seaboard, and of logboat-based extended plank sailing vessels, some of them two-masted, which could be half as big again as the basic logboat, and the strakes of which were literally sewn together with split bamboo, as were fittings and the frame elements. It should be noted that almost all of the many thousands of indigenous craft of the big rivers were shell built, that is, they were constructed by building a boat shape of planks joined together edge to edge by a process akin to an act of sculpture, without moulds, frames or models of any kind. Furthermore, the 'frame' timbers (not thought of by the builders as continuous frames in the European sense, but as two or three overlapping but not structurally joined strengthening elements shaped to fit the boat's shell) were made, fitted and fastened after the boat shape was complete, or nearly complete. Indeed, some quite large boats were never given 'frames' at all but worked as floating shells, perhaps given frame elements as they grew old and weakened by hard work for some years. In the profusion of hull forms which comprised the river boats there were at least three variations which usually proved on examination to be frameless or to have had floor timbers only. The boats were almost all built with flush-laid sides, the strakes joined in half laps by metal staples, *patam lowhar* ('leaves of iron'), and thus gave the superficial impression to the casual viewer of being skeleton-built, smooth-skinned boats on the European pattern.[4] In the 1950s, when these boats were still in their prime, the half-lap joint appeared to be universal and we never examined a boat in detail which did not show this characteristic. Even motor launches were built in this way.

The best way to learn more in detail about the Bengali boatbuilding tradition was to commission the building of a boat for our own recreational use. This we were able to do through a friendship we had formed with Lutfur Rahman, a Muslim who was something of a small entrepreneur, who had learned boatbuilding in Calcutta and migrated to Bangladesh when the subcontinent was partitioned in 1947. I have described the process of building in detail in Chapter 4 of *Boats and Boatmen of Pakistan* (Newton Abbot, 1971). Three different groups of craftsmen were employed in succession: sawyers who converted big hardwood logs into planks with pitsaws; blacksmiths who made all the fastenings for the boat from rod and bar iron in a smithy they set up under a bamboo shade cover; and finally the *mistris* (boatbuilders). To quote from *Boats and Boatmen*:

2

They marked the beginning of her construction with a simple ceremony which involved bells and garlands of flowers. Her conception was the laying of a keel plank along a row of small hardwood logs, lying like railway sleepers side by side at regular intervals. The keel plank was a little thicker than the other planks in the boat were going to be. It was held firm with bamboo pegs and with short lengths of jute cord passed through paired holes drilled through it at intervals of about $\frac{1}{2}$ ft. and pegged into the ground at either end. The *mistris* made up their jute cords themselves when they needed them, twisting the fibres of the dried jute plant by hand as they went along.

At one end of the keel plank the stem piece was attached, a massive piece of teakwood rough hewn to shape with adzes and only later cut to its final form. At the other end, a transom of heavy teak planks was cut out and joined to the keel plank. The shell of the boat was then built up plank by plank ...

There was, however, another great difference between the Bangladesh smooth-skinned boats and twentieth-century North European boat structures. The strakes of the overwhelming majority of Bangladesh boats ran in a uniform curve more or less parallel with the sheer line ending on a horizontal, or near-horizontal, line at the end, well above the water line. Usually, but by no means always, above the plank ends there were some more or less horizontal strakes which completed the structure and constituted the sheer of the boat. Where these did not exist and the strakes were brought up to almost a point at the bow, the hood ends of the strakes were fastened to a pointed *goloi*, a solid block of wood which was an essential part of the vessel's structure. Medieval iconographic material suggests a similar solution to the problem of the hood ends of the 'hulc' [a medieval ship type discussed further below; see also the Appendix].

There exist in northern Europe literally hundreds of depictions, all showing variations of this style of planking: in stone and wood carvings; on illustrated manuscripts; in jewellery decoration; in graffiti on walls and on documents; in one or two metal models made as wine neffs or censers; on at least one oil painting; and on town seals and coins dating from the early ninth to the late fifteenth century.[5] It appears that what we might now call Bangladeshi boat structures were, at least as far as the run of the planking was concerned, widely used in medieval and pre-medieval times in northern Europe and perhaps much earlier. This form of planking has been identified by Heinsius[6] and others with the vessel type referred to in numerous contemporary documents as the 'hulc'.

There appear to have been three main traditions of ship building in northern Europe in the eleventh and twelfth centuries. One was the round hull clinker tradition, one branch of which comprised the Scandinavian 'Viking' ships in all their varieties. The second, in terms of the years of its predominance, was the tradition of the cog, the flat-bottomed vessel with high clinker-built sides, from Germanic roots, which grew to be the cog of the Hanse, almost the standard merchant vessel of northern Europe for a century or more, and which, apart from its commercial importance, was the vehicle of Germanic expansion into the eastern Baltic and western Russia. The third tradition was that of the hulc, which from roots that may go back for many centuries became the preferred North European ship type in the 1300s.

Numerous archaeological finds have given us detailed knowledge of the first two boat types. Extraordinarily, despite the massive iconographic and documentary evidence for their existence over several hundred years, no identified remains of a hulc have yet been found, and indeed all we really know beyond all doubt about the hulc is the characteristic run of the strakes, ending on horizontal lines of bows and stern (or later at the bows only). The Bangladeshi structures were therefore an important discovery, undoubtedly of some relevance to the study of medieval shipping.

In this context of special interest was the discovery of vessels, usually almost without frames, massively built with strakes laid clinker fashion, that is, with the planks overlaying one another at their edges. These vessels were concentrated in the Sylhet District of northeastern Bangladesh, up against the border with Indian Assam, and appeared to exist in an almost evolutionary sequence from a small canoe with a narrow logboat as a keel to a sophisticated cargo carrier able to transport substantial weights down the many miles of river to Dacca.

These boats were the more interesting because their clinker construction showed very clearly, in varying degrees, some of the apparent characteristics of the planking of the hulc. It would appear – and this hypothesis has perhaps to a degree been confirmed by the small experiment with models to which Professor McGrail refers in the final chapter of this book – that, in a full-bodied hull planked hulc fashion, long runs of straight parallel-sided timber can be used together with very short, slightly shaped separate pieces at bow and stern. This planking pattern is very economical both in material and in labour and perhaps gives some hint of one of the factors which may have been in play in medieval, and indeed pre-medieval, European shipbuilding.

But there was more to it than that. In all of them, the clinker was 'reversed', that is, with the upper edge of the lower strake outside the lower edge of the strake above it. European clinker planking is, of course, the other way round. But the reversed clinker technique is illustrated, together with hulc-type planking, on the seal of the Admiralty Court at Bristol, established in 1446, on the seal of Amsterdam of around the 1300s and by Ian Friel in his book *The Good Ship*[7] from an English manuscript of the early twelfth century. There are no doubt other, as yet unrecognised, depictions of reversed clinker in the hulc form. We might have been excused for a degree of pleasure at what might, with sufficient imagination, have been perceived as a reincarnation in twentieth-century Bangladesh of some types of medieval hulc.

This spare-time fieldwork was conducted in Bangladesh intermittently between 1950 and 1959, usually in very difficult circumstances, with temperatures frequently in the high 90s Fahrenheit, an equally high degree of humidity, and the cyclones of spring and the monsoon rains to deal with as well. There was in those days no air-conditioning. There were for years no cross-country roads, and movement had to be by paddle steamers, some of them magnificent relics of the 1880s, by narrow-gauge steam railways, sometimes with more passengers on the roof or hanging on outside than in the carriages (which from time to time we found ourselves sharing with gigantic cockroaches), or by light aircraft of uncertain maintenance flying into abandoned wartime airstrips. It was all an adventure, but it was cut short by disaster.

The results of our labours were published in various books and papers referred to in chapters in this volume. The work was done in the mid-twentieth century, and for years it was my hope that it could be followed up by professional ethnographers with a maritime interest and expertise. And followed up not only in Bangladesh but also in other areas of the 'East' before it was too late. In fact the application of the internal combustion engine to water transport and fishing began in Scandinavia at the start of the twentieth century and was rather slowly established in Britain and North America by mid-century. It was not until after almost one hundred years that it began rapidly to develop in Bangladesh and other under-developed areas of South Asia and to transform the standard of living and eventually the whole way of life not only of those who lived and worked on the water but of the thousands whose work depended upon them.[8]

I was delighted, therefore, when my old friend and colleague Professor Seán McGrail, who after a youth spent partly as a naval fighter pilot in the Korean War had become Britain's leading archaeologist concerned with maritime aspects of the subject, told me of his plans for a series of expeditions to Bangladesh, Bengal and the east coast of India to investigate further the local boats and vessels and the life around them.

This report on those expeditions, which were conducted over a period of five years, comprises two intermingled components. One of these is essentially ethnographic. The descriptions of fieldwork on the *masula* boats (partly the result of earlier work by Eric Kentley), of the *madel paruwa* and of the hide boats of the River Kaveri are precise professional descriptions of sophisticated boats of their kind of which in the past we have had only incomplete accounts. As such they are most valuable contributions to our knowledge of three of the ways in which *Homo sapiens* has tackled the problems of providing himself with the means of travelling over water in different circumstances. These accounts also tell us of the nature of that water, of available materials with which to build floating carriers, and of societies with differing traditions of craftsmanship.

The other component of this volume comprises not only comprehensive documentation of selected 'country boats' but also important contributions towards our understanding of pre-medieval and medieval water transport with all its implications for economic and inevitably social development. I have already written of the hulc, which for several hundred years appears to have played a role, latterly an important role, in the growth of European trade and cultural expansion and yet about which we know so little. The chapter on the frame-first vessels of Tamil Nadu is a contribution to the growing study of the late and post-medieval developments in ship building which were to be of the greatest importance to European development and expansion.

The medieval 'Viking' ship, the cog and the hulc had in common that they were shell built: that is, the form of the hull was built first of planks shaped and joined together at their edges, and the strengthening internal members were then shaped to match the shell and fitted into it. These vessels were limited in size and strength by the building technique and were simply not strong enough to undertake long ocean voyages with heavy cargoes. When they attempted it, even over short sea routes such as that from Scandinavia to Iceland and on to Greenland, there were many losses. Moreover their windward sailing capacity, even that of the Viking ships which were the best windward performers of all these vessels, was limited, and many could scarcely sail to windward at all. The building of bigger and stronger vessels, capable of ocean voyaging in prolonged adverse conditions with heavy cargoes, supplies and big crews, including fighting men and eventually guns, was achieved (with, apparently, experimental fumblings) in different areas by reversing the building process so far described and first erecting a skeleton of frames and then wrapping it round with a skin of planks not joined together at the edges.

This technology, combined in the 1400s with a three-masted rig, largely of squaresails, led to what has been described as 'the invention of the sailing ship', 'the space capsule of the Renaissance' – vessels capable of prolonged ocean sailing and, providing landings could periodically be made, virtually self-supporting as far as maintenance was concerned. This was the vehicle by which in the forty years from 1480 to 1520 European seafaring expanded from the coasts of that continent and of North Africa to the circumnavigation of the world, the rounding of the Cape of Good Hope into the Indian Ocean, and the establishment of the Spanish and Portuguese nations in the American continents. There were, of course, many other factors, developments in navigation techniques, developments in financial systems on land and political changes, which made this great development possible. But without the skeleton-built, three-masted ship this expansion could not have happened. The invention of the sailing ship was one of mankind's greatest achievements.

Given the great significance of these developments, just how, when, where and why the birth of the skeleton-built ('frame-first') vessel took place became a matter of considerable historical interest. The change was not an easy one, and indeed shell construction remained the norm for smaller vessels in parts of Europe even into the first half of the twentieth century, as it did in Japan. Pure skeleton construction required that the vessel be conceived as a whole before building began. With shell construction, on the other hand, she grew under the hands of the builders without formal planning and usually in accordance with well-established local traditions; she could relatively easily be altered as she grew on the building blocks. The change to skeleton building meant a psychological and cultural revolution. No wonder the development appears to have been slow and the rate of adoption of the new technology to have varied greatly between countries and regions.

I well remember in 1947 talking with a man in his eighties who as a young man had worked at the building of a ketch called the *Hobah* at Trelew Creek, near Mylor in southern Cornwall, in 1879. He told me that they determined the keel length and proposed overall length and her maximum beam, the shape and rake of her transom counter, and that her stem should be straight and nearly vertical. A few fully assembled frames were made up around amidships and battens were then tacked from stem to transom and forward to form a nice entry and run. Many of the remaining frames were built up shaped to these battens. As my informant said, 'She was built to the natural run of the battens'. Many, many vessels must have been built in this way in communities all over the world, their 'design' partly preconceived and partly determined as the vessels were built. There were no mathematical calculations, though for bigger vessels, usually from at least the eighteenth century, a half model was first made and agreed between all concerned and the vessel more or less built to it. Nevertheless, it is rare for a surviving builder's half model to check exactly with the registered dimensions of the finished vessel, or indeed of any finished vessel, because in the process of building changes were made where difficulties arose, or because available timber supplies dictated, or because money was getting short, or for many other reasons. Draft drawings do not enter into the picture on any scale for the building of wooden ships, except in big shipyards, and the building of official vessels until the late nineteenth century. The thinking of most builders was, and remained to the end, essentially three-dimensional and this itself was possibly an ultimate legacy of the traditions of shell construction.[9]

The work by Professor McGrail's team in the East, and other work elsewhere to which he refers, tends perhaps to support the theory that at least one strong strand in the spread of frame-first construction in the late fifteenth and in the sixteenth century came from the Iberian peninsula and was carried overseas by Portuguese and Spanish voyagers. The relics of it may survive in Tamil Nadu, as elsewhere in the areas of the East and the West penetrated by the Iberian venturers. If the Iberian peninsula is one of the areas in which frame-first, skeleton construction as we know it began its long gestation, then it is likely that it spread from here into northern Europe. But the migrations of the technique were undoubtedly complex and its local development followed various routes to the adoption of full skeleton construction.

The subject of these reports is, of course, multi-dimensional. The heart and object of the work may have been the study of the construction of the boats, but Professor McGrail and his colleagues deal also with the social context and the handling of the boats and their functions, crew duties, fishing methods, and the way the men sometimes regard the boats. The crew of the *vattai* thought of their vessel as 'almost a spirit'. The development of a special relationship between men and the boat on which they depend for a living is by no means confined to the East. The use of the female 'she' to refer to a boat, rather than the neuter form, is to be found in many cultures. In Swedish, a boat can be referred to simply by her name, say *Svan* (Swan). But sometimes it is likely she will be referred to as *Svanan*, the female swan, a term indicating a special relationship with the boat, as if she was animate. When, in the late 1930s, the now aged ketch *Hobah* (the mysteries of the building of which I probed in the 1940s) was coming to the end of her working life, Captain William Lamey, the son of her owner, Charles Lamey, would sometimes find his father sitting on the village quay near where the vessel was moored. When asked to come to supper, he once said: 'Boy, I'm just looking at the *Hobah*, just looking at the *Hobah*. She gives me a kind of peace.'

NOTES

1 *Youth* was published first in 1902. The quotation is taken from p. 43 of the 1965 edition.
2 Conrad, when he was writing a story closely based on his own experience, naturally romanticised to a degree. A factual account of the *Palestine*'s passage to the East is given in Jerry Allan, *The Sea Years of Joseph Conrad* (London, 1967): 153–60.
3 See, for example, W.H. Coates, *The Old Country Trade* (London, 1911).
4 This building process is described in detail in B. Greenhill, *Boats and Boatmen of Pakistan* (Newton Abbot, 1971), Chapters 4 and 5. Building as 'an act of sculpture' without moulds of any kind was, of course, widely practised in northern Europe right through to the end of the twentieth century. In *Maritime Life and Traditions* (2000), No. 8, pp. 9 and 10, Gloria Wilson writes of Whitby in Yorkshire as late as the 1990s:

> In the 1980s and early 1990s C.A. Goodall was the principal local coble builder producing about a dozen boats from 28 feet to 37 feet long. Yard owner and master boatbuilder Tony Goodall, who retired in 1995, told me that he used no moulds when planking a coble. Hull shape was created by using a known set of plank widths and bevel angles for the plank lands at stem, shoulder, midships, tunnel and stern.

5 For a more detailed study of these structures and their origin, see B. Greenhill, 'The Mysterious Hulc', *Mariner's Mirror* 86, no. 1 (February 2000): 3–18.

6 Heinsius, Paul, *Das Schiff der hansischen Frühzeit*, Hanseatic Historical Society, N.S., vol. XII, published by Herman Bolar's successors (Weimar, 1956).

7 Friel, I. *The Good Ship* (London, 1995): 37, figs 2–4.

8 For a very well documented and powerfully written study of the effects of modernisation on a maritime industry, the Canadian Banks fishery, and the livings of those concerned with it in the mid-twentieth century, see R. Andersen, *Voyage to the Grand Banks* (St Johns, Newfoundland, 1998).

9 See Greenhill, B. and Manning, S., *The Evolution of the Wooden Ship* (London and New York, 1988).

1

AIMS AND METHODS

Seán McGrail

The aim of this monograph is to describe selected boats of South Asia – their building and their use – as they were at the end of the twentieth century. The four principal chapters (Chapters 2, 3, 7 and 8) are based on fieldwork undertaken by a multi-disciplinary team between 1996 and 2000. Two members of this team are maritime archaeologists (SMcG and LB), another is a maritime ethnographer (EK) and the fourth is a naval architect (CP). This fieldwork and the subsequent research were financed as a project of the Society for South Asian Studies under the title 'Boats of South Asia'.

During successive winters, periods of three to four weeks' duration were spent on the Bay of Bengal coast of India: in Orissa, in early 1996 (Chapter 3); in Tamil Nadu, in early 1997, in 1999 and 2000 (Chapter 7); and in West Bengal, in late 1997 (Chapter 3). Fieldwork was also undertaken in late 1997 on the headwaters of the River Meghna in the Sylhet District of Bangladesh (Chapter 2); and in 2000 a brief period was spent on the headwaters of the River Kaveri in northwestern Tamil Nadu (Chapter 8).

In addition to these four chapters with joint authorship there are four single-author chapters. Chapters 5 and 6, on sewn plank boats, are based on fieldwork undertaken by Dr Eric Kentley for the National Maritime Museum during 1983, 1984 and 1986 in Sri Lanka and eastern India (Orissa, Andhra Pradesh and Tamil Nadu). Chapter 4 is based on Colin Palmer's investigations in the delta region of Bangladesh between 1985 and 1994 when he was working for the Bangladesh Inland Water Transport Authority. In Chapter 9, Colin Palmer gives theoretical evaluations of the Bangladesh *barki*, the Orissan *patia* and the Tamil *vattai* – the three plank boats dealt with in greatest detail in this volume – as well as the *vallam*.

One outcome of our research in South Asia has been an increasing awareness of how the documentation of traditional boats can lead to a greater understanding, not only of South Asia's maritime past but also of aspects of European maritime archaeology and history. An example of this is given in Chapter 7, which is concerned with Tamil frame-first boats and ships and the possible European origins of the methods used to design them. There is also a European aspect to Chapters 2 and 3, which deal with the reverse-clinker boats of Bangladesh and Orissa/West Bengal. In this case a direct link between South Asia and Europe is not proposed, rather the application of South Asian analogous evidence to a late-medieval European problem: how reverse-clinker planking and/or hulc planking patterns might be recognised archaeologically. The European aspects of this topic are discussed in the Appendix.

This present chapter, describing our aims and our methods, is followed by descriptive chapters ordered, more or less geographically, from Bangladesh in the northeast to Sri Lanka and Tamil Nadu in the south. These seven chapters are followed by a naval architectural assessment of key boat types and, in a concluding chapter, the work done by the Boats of South Asia project is evaluated and future lines of research are suggested.

The Boats of South Asia project

In 1994 I was invited to India by the National Institute of Science, Technology and Development Studies (NISTADS) to play a part in a maritime archaeology training course in Tamil Nadu. With a travel grant from the Society of South Asian Studies I visited Tamil University, Thanjavur, where I lectured on maritime archaeology and ethnography and accompanied selected students on a field trip, examining rafts, boats and boatyards along the coast of southern Tamil Nadu as far south as Cape Comorin/Kanyakumari (McGrail, 1995A). I also learned about the university's research programme concerning the maritime history of southern India from Professor Victor Rajamanickam, head of the Department of Earth Sciences.

As a consequence of that visit and the subsequent Delhi conference on 'Techno-archaeological perspectives on seafaring in the Indian Ocean' (Ray and Salles, 1996), it became clear to me that with little, if any, conservation infrastructure and virtually no personnel trained in the location, excavation and interpretation of boats and landing places, fieldwork into those aspects of maritime archaeology could not, at present, be undertaken in India or, indeed, anywhere in South Asia. On the other hand, it was equally clear that the documentation of today's traditional working boats and rafts could be undertaken with relatively little capital outlay. Indeed, such ethnographic work was crying out to be done; although much had been learnt earlier in the twentieth century about indigenous boats in certain regions of South Asia (Hornell, 1946; Greenhill, 1971; Deloche, 1994), there had not been any serious and systematic attempts to investigate the water transport of vast regions of the subcontinent. Furthermore, if Indian scholars could become involved in the recording of traditional water transport, a group of people might well be gathered together who would be enthusiastic about South Asia's maritime past and would have the appropriate knowledge and practical skills, which they could use at some future time when the location and excavation of ancient rafts and boats might become a real possibility.

11

Using such a long-term approach, the goals of boat archaeology might be achieved via boat ethnography.

The Society for South Asian Studies subsequently awarded me a five-year research grant to undertake the study of selected boats of South Asia. The principal aim of this project was to document traditional South Asian boats – those propelled by wind and/or muscle power – in two complementary ways:

a) in their own right as a record of a way of life which was fast disappearing as engines were fitted, and as metal and plastic replaced wood; and
b) to throw light on South Asia's maritime past and, whenever possible, on maritime affairs elsewhere and at other times.

It was also an aim of the project to encourage South Asian scholars to undertake similar work, since the field is vast and time short. The documentation of a boat type, its building and its use is an art and a science which has very few practitioners anywhere in the world, and none were known in South Asia.

Selection of the boat types to be documented

It was decided to concentrate research on the east coast of South Asia for two main reasons:

a) One of the team (EK) had undertaken several seasons of fieldwork on that coast, and another (SMcG) had had recent experience there.
b) It was evident from earlier studies that there had been less outside influence on boatbuilding techniques there than on the west coast.

Within this region, specific traditional boat types were selected for documentation, not only because they constituted an important regional tradition but also because they had features that were, in our experience, unusual and because they appeared likely to illuminate studies in maritime history and archaeology, both within and without South Asia. Three such traditions were selected.

The reverse-clinker tradition

Professor K.S. Behera of Utkal University, Bhubaneswar, had reported to the Delhi conference that fishing boats with reverse-clinker planking were used in large numbers off the northern coast of Orissa. This type of planking was known to be very rare worldwide, but there was evidence that it might have been used in medieval northern Europe. This boat type thus met our threefold criteria. The documentation of this Orissan *patia*, a reverse-clinker sailing boat, subsequently led to the study of a closely related oared boat in nearby West Bengal.

Dr Kentley and I had been on the staff of the National Maritime Museum, Greenwich, during the later years of Dr Basil Greenhill's directorship and we were familiar with his pioneering study of Bangladesh reverse-clinker boats. It was natural therefore to extend our research further north, to Sylhet in northeastern Bangladesh, to document the *barki* and other boats similar to the ones Dr Greenhill had recorded almost fifty years earlier.

A frame-first tradition

In 1994 I had noted several Tamil ships and boats that had been built frame first: that is, the frames were first fashioned and assembled to give the shape of hull required, then the planking was fastened to the framing. This was radically different from all other known traditional vessels on the Bay of Bengal coast, which were built plank-first, the framing being added to the planked hull. Thus the second area of investigation was determined: this tradition was clearly well worth documenting in detail. Moreover it seemed likely that documentation of these vessels would throw light on an important period in the technological history of late-medieval Europe when frame-first ships began to be designed and built in increasing numbers in the Mediterranean and in Atlantic Europe.

These distinctive frame-first vessels in coastal Tamil Nadu ranged in size from small fishing boats some 5 m in length to ships of c. 38 m and 650 tonnes' capacity. It was decided to document a 12 m *vattai* fishing boat, as representative of this tradition. A second season of fieldwork was subsequently spent in coastal Tamil Nadu, recording variations observed in other boats of this tradition, especially in the design process.

A hide boat tradition

The hide boat has a long history of use worldwide, at sea and on rivers and lakes (McGrail, 2001A). In southern India such boats have been used on rivers since at least medieval times (Deloche, 1994: 138), yet they have never been recorded in detail. A fleeting opportunity was therefore seized to document one of the many *parical* hide boats still in use on the River Kaveri, at Hogenakal in northwestern Tamil Nadu, close to the border with Karnataka.

A training course

Our association with Tamil University led, in our final season, to a training course there under the auspices of the Department of Earth Sciences. The head of that department, Professor G.V. Rajamanickam, had selected nine Indian archaeologists and historians from the Archaeological Survey of India, the National Institute of Oceanography in Goa, and from universities and colleges in other states. By means of lectures, fieldwork, practice interviews and practical measurement they were instructed in the elements of documenting a small fishing boat (McGrail, 2001B). In this way we went some way towards achieving the subsidiary aim of the Boats of South Asia project: to involve South Asian scholars in this type of research.

Maritime archaeology in South Asia

There is clear evidence for the early use of South Asian rivers and also for overseas trade (Chaudhuri, 1985: 1–162; Ray, 1993, 1994). However, only one excavated craft is known in the whole of the subcontinent: a logboat from the Kelani Ganga in the Colombo district of Sri Lanka, which is dated to the sixth/fourth centuries BC (Weizmann Institute, Jerusalem; Vitharana, 1992; Davendra, 1995).

The Archaeological Survey of India has not yet extended its research to sites in estuaries and former river channels where ancient boats might be found, and few professional archaeologists in India have excavation experience in the maritime zone. The situation elsewhere in South Asia is understood to be similar. The National Institute of Oceanography in Goa has undertaken underwater excavations, but these have mainly been on the sites of submerged harbours and post-medieval wrecks. This minimum involvement policy has probably been wise, since there are only very limited facilities for the conservation of waterlogged wood, and there is no institution specialising in the interpretation and display of ancient boat remains.

Early South Asian water transport

What little is known about early water transport has come from documentary and iconographic evidence – in this respect, South Asia is no different from most countries.

Rafts and boats of the Indian subcontinent are mentioned by name (e.g. *sangara*, a log raft) in the *Periplus Maris Erythraei* of the first century AD (Casson, 1989), but structural details and methods of propulsion and steering are scarcely noted. Water transport is mentioned in similarly generalised terms in Buddhist and Jaina texts and in other early Indian sources.

There is also some representational evidence: depictions on seal, amulet and potsherd of c. 2000 BC from Mohenjo-daro, and there are clay model boats from Lothal of about the same date. Vessels with a mast are depicted on coins, seals and monuments of the second/first century BC from sites in the Ganges delta and others as far south as Sri Lanka. The vessels depicted on a medallion of the second century BC from the monastery at Bharhat and on the east gate of Stupa 1 at Sanchi have planking which appears to be fastened together by double-dovetail wooden clamps (McGrail, 2001A: 253–5).

From the fourth to the sixth century AD there is further evidence. Coins and sealings from the Andhra coast show two-masted ships with two steering oars; other vessels are depicted on coins from the Coromandel coast. There are also depictions in Buddhist places of worship at Aurangabad and Ajanta (Ray, 1986: 117–20; Ray, 1994: 162–88). The sum of this evidence contributes to our knowledge of early vessels in South Asian waters but does not provide the detail needed for an understanding of early boatbuilding and seafaring.

There is then an apparent gap in the evidence for water transport until the sixteenth century.

Maritime ethnography in South Asia

Ethnography, formerly a sub-discipline of anthropology, may be defined as 'the description and analysis of the material and social aspects of recent or present-day, non-industrial, small-scale societies': recording boats was once within its domain. During the sixteenth to the eighteenth century European explorers, traders, missionaries and seafarers published brief descriptions of some of the boats and rafts of South Asia (Hill, 1958): they were, in effect, proto-ethnographers.

In the nineteenth and twentieth centuries, accounts and drawings of South Asian vessels were compiled by a number of well-informed observers. These included John Edye (1834), the master shipwright of Trincomalee Naval Yard; the French naval officer Lieutenant (later Admiral) F.E. Pâris (1843); and the marine biologist James Hornell (1920; 1924; 1926; 1930; 1933; 1943; 1946 and 1970). By the time that Hornell's major work, *Water Transport*, was published in 1946, the study of the material aspects of twentieth-century societies was already becoming unfashionable in anthropological circles and attention was increasingly focused on social relationships. Towards the end of the twentieth century, however, ethnography began to reappear in universities under the guise of 'material anthropology' or as 'ethno-archaeology'.

In South Asia this mid-twentieth-century decline in the popularity of ethnography meant that, for a quarter of a century or so, boats were little documented. In more recent times, however, notable contributions to water transport studies of the Indian subcontinent have been made by Greenhill (1957, 1961, 1971), Kentley (1985, 1993, 1996), Kentley and Gunaratne (1987) and Kapitan (1987–9), while Deloche (1994) has published a concise historical account of the boats and rafts of India's coastal regions, together with geomorphological and environmental descriptions of coastal and estuarine waters.

Ethnographic accounts from the sixteenth to the twentieth century have certainly increased our knowledge of water transport. For example, if we take broad categories of vessels still in use in South Asia today and use all forms of evidence available, we can trace them back in time (McGrail, 2001A: Ch. 6):

- log rafts to the late seventeenth century (they were first noted in the first-century *Periplus* [Casson, 1989: Ch. 60]);
- logboats to the early seventeenth century;
- hide boats to the late fourteenth century;
- sewn plank boats to the sixteenth century;
- boats with reverse-clinker planking to the late eighteenth century; and
- frame-first vessels to the early twentieth century

The evidence on which this summary is based is very sparse, however, and in many cases structural details have not been recorded. Clearly, in preparation for the time when it may become practicable to excavate, there is much work to be done on documents, inscriptions and representations to learn more about South Asia's maritime past.

There is also an urgent requirement for today's water transport to be recorded in detail. Deloche (1994: 180, fig. 37) has published a map based on a sketch by Sopher (1965: 12, 16) which shows the mid-twentieth-century distribution of certain broad categories of water transport (Fig. 1.1). This is very informative, especially when studied in conjunction with Deloche's descriptions of the appropriate coastal waters. However, since there have been no systematic surveys of South Asia's water transport such a map had to be based on a very few reports of varying standard. Inevitably, therefore, the map is a generalisation. The detailed distribution of the various categories of water transport (at least on the Bay of Bengal coast) appears to be more complex. Systematic and extensive documentation of all traditional rafts and boats will have to be undertaken before a detailed distribution can be plotted.

Documentation of water transport in the twenty-first century

In comparison to an archaeologist studying an excavated boat, the recorder of traditional boats has several advantages, not least that the representative nature (or otherwise) of a particular boat can be established, that the boat structure is generally complete, and that the uses of the boat can be established. A further great advantage is that the users of the boat, and often also the builder, can be questioned. Questions to be asked include: How do you handle the sails when tacking/wearing? How deep do you load the boat? What types of fish do you catch? How do you keep your reckoning when out of sight of land? What is the function of that notch on the inner face of the stem? Why was this boat given that shape of bow? Why was reverse-clinker planking used and not some other form of planking? Thus an experienced

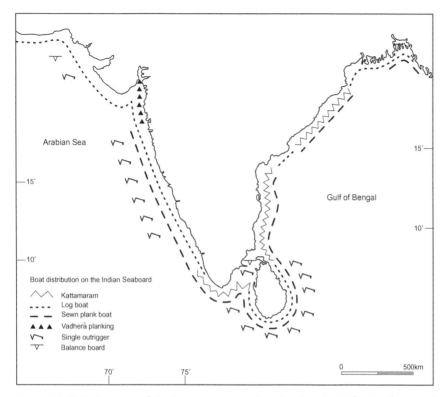

Figure 1.1 Deloche's map of the South Asian coastline showing the 1965 distribution of log rafts, logboats and sewn plank boats; and three characteristic features. After Deloche, 1994: fig. 37.

recorder can build up a near-comprehensive picture of the building of the boat, from felling the trees to launching, and of her uses and operational performance. This is a picture which archaeologists working within their own domain perceive only dimly; but this perception can often be improved by using comparative data derived from traditional rafts and boats.

The prime reason why the documentation of traditional South Asian rafts and boats should be undertaken today is so that a way of life, and a technology that is fast disappearing, will be recorded. The traditional wooden boat, propelled by wind or by muscle power, is being replaced by metal or plastic boats with engines. The traditional log raft is also being phased out. By recording the building and use of such craft, we preserve evidence of these skills for posterity.

From the historical and archaeological viewpoints there are a number of further reasons for recording traditional boats, and these are especially relevant to South Asia. In that subcontinent there is little historical, and even less archaeological, evidence for ancient water transport, but there are

17

still small-scale, essentially non-industrial, communities building and using rafts and boats of a form and structure which suggest that they are examples of 'living traditions' surviving from much earlier times. The archaeologically orientated reasons for recording the boats of such communities are:

a) The documentation of traditional water transport provides a baseline and a launch pad for historical research backwards from today.

b) Excavators can be prepared to encounter fittings and features which are generally found associated together in traditional boat types.

c) Evidence from traditional boats can be useful in the identification of wear marks and unusual fittings on excavated boats.

d) Such evidence can also be used, when appropriate, in the theoretical reconstruction of the missing parts of excavated boats. However, archaeologists should also learn from the study of traditional boats that a particular aim in boatbuilding can be achieved in several ways. For example, the ends of a plank boat can be closed and made watertight in a number of ways. The problem is universal; solutions differ. There may not be a unique solution to an archaeological reconstruction problem.

e) The archaeologist becomes familiar with the work – indeed, the way of life – of seafaring and riverine communities, which cannot but increase understanding of similar communities in earlier times.

Behind several of these reasons lies the idea of cross-cultural comparisons, and the question must be asked: are such comparisons valid? The more alike the two cultures (one ancient, one modern) are, in environmental, technological and economic terms, the greater the likelihood that cross-cultural comparisons will help in the reconstruction of ancient technologies. However, such analogies over time and space, like all forms of evidence, need to be treated with caution, even within one country. The fact that log rafts and sewn plank boats are in use in southern India today does not prove that they were used in India 3000 or even 300 years ago. It does, however, draw our attention to such a possibility (even in the absence of excavated evidence) and impels us to search historical accounts and representational evidence to see whether we can trace such use in earlier times.

Recording a traditional boat

The standard to be aimed at when recording a traditional boat is the same as in a boat excavation, i.e. to compile a record from which a competent model builder could build an accurate model and from which a detailed account of the boat's routine uses could be written.

Having defined the tradition of boat that is to be documented, a representative example (an archetype) has to be identified. Agreement for the use of this boat for the required period must be reached with the owner and the boat moved, if necessary, to a suitable site above high water. The boat is then levelled and a centreline rigged.

The hull

A hull reference system is established. All planks, timbers and fittings are identified and given a codename: for example, strakes may be labelled P1 to 'Pn' to port and S1 to 'Sn' to starboard, from the keel or foundation plank outwards. Framing timbers may be similarly labelled F1 to 'Fn' and beams B1 to 'Bn', from bow to stern. General and context photographs are then taken and notes compiled on the main features of the boat, including overall dimensions. References are made in these notes and in the photographic record to codenames.

A boat can be documented by one, two or three people. It is possible for one person to measure and compile a drawing on site, but such a detailed, measured drawing can take a very long time. More usually, only selected measurements are recorded, including at least one transverse section. These are subsequently used off site, in conjunction with notes and photographs, to compile a basic measured drawing.

With two people, one measuring and one drawing, a measured drawing can readily be done on site. A third person speeds up the documentation, since photography and direct measurement can be undertaken at the same time as the measuring. This third person can also concentrate on establishing the 'internal stratigraphy' of the boat so that the sequence of building can be deduced.

The measured drawing

A measured drawing of a boat's hull can be compiled in several ways: we prefer to use offsets from horizontal and vertical datum lines to produce a plan, several transverse sections and a longitudinal section – as on an archaeological excavation (Fig. 1.2). For boats up to c. 12 metres in length, a scale of 1:10 results in a drawing of manageable size, but still with features well defined.

a) For the plan: horizontal offsets can be taken either from a centreline datum (preferred) or from parallel datum lines along both sides of the boat.

b) For the longitudinal section: vertical offsets are taken from the centreline datum.

c) The stations for the transverse sections are marked along the centreline, and vertical offsets are taken from transverse datum lines at these stations. Where the shape of the boat is changing rapidly, as towards the

SETTING UP THE CENTRELINE DATUM

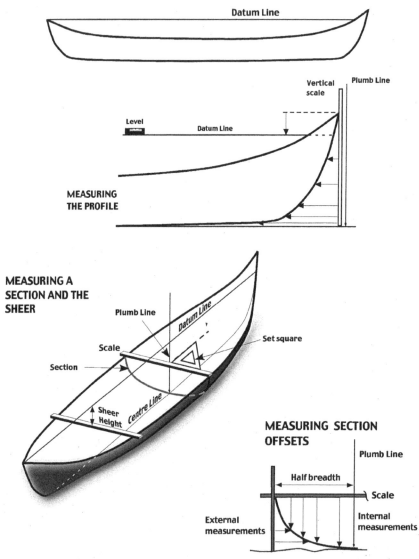

Figure 1.2 Diagram to illustrate the standard method of recording information for a measured drawing of a traditional boat.

bows, stations should be closer together than elsewhere. Some of these sections should be at frame stations and some in between. Generally speaking, boats are symmetrical about their middle line, so only a little more than half of each section need be recorded, unless there are asymmetric fittings or features.

d) For profiles of the outer contours of a boat other appropriate offsets are taken.

e) The sheerline is recorded by vertical offsets from the centreline datum to a straight edge laid across the boat, from sheer to sheer, at intervals along the length.

f) Stem and stern posts are measured by horizontal offsets from a vertical datum at the ends. The posts' outer profile, and an inner profile where the strakes end, can also be drawn in this way.

Direct measurement

Direct measurements have to be recorded of the dimensions of those features and fittings that are especially important: thickness of planking; moulded and sided dimensions of key timbers; spacings (centre to centre) of fastenings and of frames. Scarfs and other joints, and details of fastening methods, are best recorded in a diagram with measurements.

Propulsion and steering

Certain aspects of the propulsion and steering arrangements will appear on the hull drawing: for example the mast step, and the pivots for oars and for any steering oar. It will also be necessary to draw and photograph mast, yard, sail, rudder and oars. Diagrams with measurements should be made of the rowing geometry and of the rigging.

Equipment

Anchors, bailers and fishing equipment, unless of special interest, can be recorded by photographs (with scales) and in technical notes.

Photography

It is impossible to overemphasise the importance of the photographic record: colour, black-and-white and slides. General and detailed shots should be taken; scales, an arrow pointing towards the bows, and the codenames should be in the field of view whenever practicable. This archive will prove invaluable later when preparing a description and a publishable drawing of the boat. A video of a boatbuilder at work can help resolve problems encountered when compiling the detailed sequence of construction.

21

The context of the boat

Features of the landing place from which the boat generally operates should be documented: landward access; nature of beach; aspect; predominant winds; tidal pattern; offshore hazards etc. A sea passage in the boat recorded, or a similar one, is highly desirable, even if it is only a demonstration and not a working sortie. The way the boat is handled and other aspects of life on the water should be observed. These observations can then form the basis for an interview with the crew after returning to the landing place. The many other questions on which the crew may throw light can be deduced from accounts given in subsequent chapters of this book.

At least one builder (but preferably several) of the type of boat being documented should also be interviewed. The aim should be to discuss with him the design of the boat and the building process, from obtaining raw material to launching and fitting out. His boatyard should also be studied carefully since there will be clues everywhere (not least his tools) to the techniques he uses. Moreover, incomplete boats will reveal details of hull structure that cannot be seen on a working boat.

Time taken to record

A *parical* hide boat was documented in less than three hours in Tamil Nadu (see Chapter 8). This was a very simple boat, however, recorded during an unexpected interval in fieldwork. Experience suggests that, because of unavoidable logistical problems, the lack of shade on the foreshore (even in winter) and the inevitable difficulties of having to work through an interpreter (probably with little, if any, nautical knowledge), a two-man team should generally allow five days to document an open boat in the tropics. A model schedule would be:

Day 1   Reconnaissance
Day 2   Record
Day 3   Complete recording; compile technical notes; complete photo-graphic record.
Day 4   Draw fair version of plans and sections to see what is missing and to note anomalies. Write brief description of the boat and note any gaps in the narrative.
Day 5   Return to the site to record missing data and resolve anomalies.

A three-man team might take four, or possibly three and a half, days rather than five. Clearly, a major problem here is how to ensure the boat is not moved during this protracted documentation. This is where photo modelling and other innovative recording methods should have a distinct advantage.

The sea passage, the interviews with the boat's crew, one visit to a boatyard and the interview with one builder can take another two days, or one day if the team splits into two groups. Dividing into groups is seldom desirable, however, since extra auditors to an interview provide much-needed cross-checks when an account of what was said is being compiled subsequently.

Once all this has been done, the examination of similar and related boats for variant features and further interviews need not take more than three or four days, depending on how widespread the survey is to be. This wider enquiry is probably best done in the following season of fieldwork, when the knowledge gained in the first survey has been assimilated, a draft of the report has been written and a fair drawing of the archetype boat has been compiled.

Future recording methods

The advent of computer-based data processing and laser-based surveying equipment holds the potential to provide rapid and less demanding and disruptive alternatives to traditional methods of recording working boats on the foreshore. One approach uses photographic images from which three-dimensional data are deduced by computer-based processing. The other relies on direct on-site measurements of three-dimensional space using a surveying laser system.

Both these approaches have the great advantage that it is not necessary for the boat to be level or with clear access all around. The photographic method in particular can be completed very rapidly in the field (although the subsequent data processing can be lengthy).

In the photographic process, a number of pictures are taken of the boat, ideally from ahead and astern, on the quarters and from abeam. Each photograph should contain a known dimension (which can be a dimensioned scale or two points marked with chalk and measured by tape). The images (most conveniently taken on a digital camera) are then read into a computer and processed using a program such as Photomodeler. This technique is most useful for the production of a lines plan but can become very tedious indeed if all the structural details are required.

The method using laser survey equipment to record boats in their operating environment requires more bulky equipment but offers the possibility of high accuracy and the capability to feed data directly into a computer-based 3D modelling package. For boats in museums, laser scanning techniques can be used (Moreton *et al.*, 2000).

Further reading on documenting ships and boats

Blake, W.M., 1935. Taking off the lines of a boat, *Mariner's Mirror* 21: 5–13.

Farrer, A.P., 1996. Recording a craft's lines, *Mariner's Mirror* 82: 216–22.

Lipke, P. *et al.* (eds), 1993. *Boats: A Manual for their Documentation*. Nashville, Tennessee: American Association for State and Local History.

MacKean, H.R., 1980. *Catching a Line*. St Stephen: New Brunswick Heritage Publications.

Milne, G. *et al.*, 1998. *Nautical Archaeology on the Foreshore*: 51–67. London: Royal Commission for Historical Monuments (England).

2

THE REVERSE-CLINKER BOATS
OF BANGLADESH

Seán McGrail and Lucy Blue

Bangladesh has some 24,000 km of rivers (about 290 in number), streams and canals that together form one of the most complex systems of inland waterways in the world (Fig. 2.1). The three main rivers are the Padma (Ganges), the Jumuna or Jamuna (Brahmaputra) and the Meghna. Together they form a dense network: nowhere in Bangladesh is more than a few kilometres from a navigable waterway. The Padma, Jumuna and Meghna meet at a confluence to the south of Dhaka and form the world's largest delta, covering some 60,000 sq. km.

The alluvial river plains that constitute 90 per cent of the country are very flat and never rise more than 10 m above mean sea level. Less than 1 per cent of the country has an elevation of more than 75 m above mean sea level, while about half of the country is no higher than 3 m above mean sea level. The low-lying, deltaic nature of the country, the presence of the three large river systems and the heavy monsoon rainfall all contribute to Bangladesh being extremely susceptible to flooding annually. In the dry season only about 7 per cent of the country's surface is covered in water; during the monsoon period from June to September, however, these waterways swell to inundate up to half of the country (Jansen and Bolstad, 1992: 7, 9). Thus boats are, and always have been, the primary means of transport in Bangladesh.

From the historical viewpoint, river use and nautical technology in Bangladesh cannot be separated from those in the eastern parts of West Bengal, which is dependent on, and very much influenced by, the River Huegli. That river is effectively the westernmost arm of the great three-river delta.

Bangladesh weather

The dry (or cold) season in Bangladesh runs from mid-October or beginning of November to the end of January or beginning of February. During this period moderate northerly winds predominate and the rivers are relatively

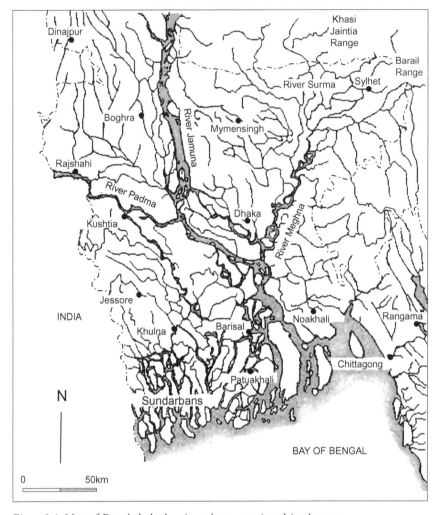

Figure 2.1 Map of Bangladesh showing places mentioned in the text.

tranquil. By mid-March the winds have backed to the south and are stronger, with occasional northwesterly squalls in the east of the region. Increased melt-wash and rainfall in the mountains cause the rivers in the north of the country to begin rising by the end of April or beginning of May, and they are in full flood by July (Rennell, 1793: 390).

The rainy or monsoon (wet) season begins in mid-June, with the onset of SW winds, and lasts until September. The rivers reach their highest levels in July/August, due to the combined effect of monsoon rains and flood waters, and begin to retreat in August/September. The downstream currents are strongest in August when the SW (upstream) winds are still dominant: thus this month is the optimum time for a round trip, using the current to assist the passage downstream and sailing upstream with the SW wind (Jansen *et al.*, 1989: 153).

Sylhet District (Fig. 2.2)

The area where fieldwork was undertaken during December 1997, Sylhet District in the northeast of the country, is a flat, low, alluvial plain with lakes and marshes. Geologically this region is a low-lying depression known as a hinge zone: a tectonically active subduction zone on the Austral-Indian plate to the south and southwest of the Eurasian plate. Sylhet District is dominated by the Khasi Jaintia hills to the north and the Barail range to the northeast. These mountains, the southernmost reaches of the Himalayas, were formed in the Tertiary period as part of the uplifting Eurasian plate and are still rising: that is, the Sylhet plain is still subsiding. The complex nature of these major plate boundaries make the district seismically very active.

The rivers Surma and Kusiyara, the two main headwaters of the River Meghna, both rise in India and, between them, drain the whole of Sylhet District. The River Surma flows in a westerly direction across the Sylhet plain through Sylhet town and then northwesterly to Chhatak, where it is joined by the River Bagra from the north and the River Piyain from the northeast. From Chattak, the Surma flows eastwards to Sunamganj, in the neighbouring district, where it turns southwards. The River Kusiyara flows roughly parallel to the Surma but further south. The two rivers join to form the River Meghna which, further downstream, becomes at least 8 km wide in the wet season.

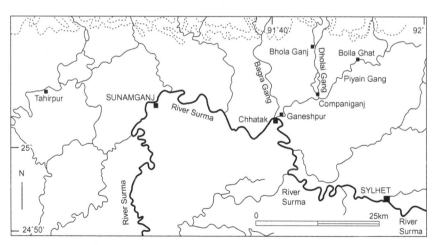

Figure 2.2 Map of Sylhet District in NE Bangladesh showing sites where reverse-clinker boats were found in 1997.

Reverse-clinker planking

Characteristics

Boats are said to have reverse-clinker planking (in Bangladesh Bengali, *digekata*) when each succeeding strake of planking overlaps *in*board the upper edge of the strake below, rather than overlaps *out*board, as is generally found in European clinker-built boats (Fig. 2.3). The strakes are fastened together, in both types of clinker work, through the overlap: in Europe by clenched nails; in South Asian reverse clinker either by hooked nails (in Orissa/West Bengal – see Fig. 2.3) or by boatbuilders' staples (in Bangladesh – see Fig. 2.18). When using the term 'European clinker' to describe the planking lap shown in the upper part of Figure 2.3, there is no implication that the clinker planking of this style that is found in South Asia (Deloche, 1994: 168–70) is European in origin.

There are no excavated examples of reverse-clinker boats in South Asia; there is, however, some iconographic evidence from the twelfth century AD, as well as documentary and illustrative evidence dating from the seventeenth century to the present day (pages 32–38, 71–72).

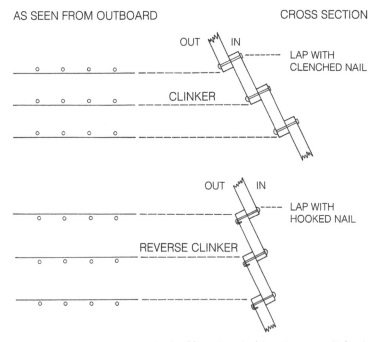

Figure 2.3 Diagram showing two methods of fastening planking: European clinker (upper); reverse-clinker (lower). Not to scale. Drawing: Institute of Archaeology, Oxford.

On three-dimensional sculptures and seals, and on the very rare examples of cross-sectional drawings that have been published, reverse-clinker planking can readily be recognised by the direction of the overlap. It can also be recognised on profile and elevation drawings providing that the strake edges are clearly delineated and there is either shading to denote the overlap or a run of fastenings along the seams. In such depictions, the two forms of clinker can be differentiated by the relationship between the fastening nails and/or the shading to these seams. With reverse clinker the fastenings are below the depicted seams, near the upper edge of each strake (Figs 2.3, 2.4), while the shading is above the seams, near the lower edge; depictions of the European form show the converse. The exception to this general rule is that illustrations of boatbuilders' staples, which cross the seams, may give no guidance; however, associated shading may indicate which form of clinker is being represented. It has to be borne in mind when attempting to interpret such depictions that a particular artist/craftsman may not have realised the technological importance of depicting the precise position of fastenings. Furthermore, some artists (for example, Solvyns in the late eighteenth century) did not, at times, depict any fastenings at all (Fig. 3.4).

Planking patterns

A feature of boats built with reverse-clinker planking today in South Asia is that their lower strakes rise towards the ends of the vessel, so that they

Figure 2.4 The fifteenth-century seal of the Admiralty Court of Bristol, depicting a ship with reverse-clinker planking. Copyright: Basil Greenhill.

29

terminate not at a post but on a level or angled surface, well above the waterline. In small boats this feature may be vestigial and difficult to discern, nevertheless it is usually there. In this configuration, the garboard or first side strake runs almost the full curved length of the vessel, from high on one end, along the keel or bottom plank, to high on the other end. All other strakes of this nature are shorter in length and do not approach the hog/false post or block end. Further strakes running the full length of the vessel may be fitted above these lower strakes, thus, as it were, sealing off the ends of the lower reverse-clinker strakes. These additional strakes may stand on the highest reverse-clinker strake, or they may overlap it in reverse-clinker or European clinker fashion.

Such planking patterns will be referred to in this volume as 'hulc planking', since this arrangement seems to be a defining characteristic of the medieval European cargo ship known as a hulc (Figs 2.5, 3.8; Greenhill, 1995A: 250–53; 2000). This planking pattern is readily recognisable both in actual vessels (Fig. 2.6) and in representations (Fig. 2.4 – bow only).

However, there are other types of boat in South Asia that have this plank pattern yet are not built with reverse-clinker planking (Fig. 2.7). The *chhoat* and the *salti* of Orissan/Bengali waters (Mohapatra, 1983: 1, figs 1, 2, 8 in appendix) and some of the Bangladesh river cargo boats known as *pallar* and *patam* (Greenhill, 1971: figs 9, 10, 11) have hulc planking, surmounted by one strake (*chhoat*, *salti* and *patam*) or by several strakes (*pallar*) which run(s)

Figure 2.5 Scene from the life of St Nicholas on the late-twelfth-century font in Zedelgem, Belgium. The ship has hulc planking. Photo: Jacques Haers S.J.

30

Figure 2.6 A large Bangladesh reverse-clinker cargo boat discharging a cargo of cement in the 1950s. The lower planking ends, not at the block stem (*goloi*), but on an angled/near-horizontal line. Copyright: Basil Greenhill.

Figure 2.7 A *pallar* under repair in the 1950s. The lower hull has 'hulc planking' which is capped by the horizontal strakes of the upper hull. Copyright: Basil Greenhill.

the full length of the vessel. Yet, externally and internally, these four types of boat are smooth-skinned and certainly not reverse clinker (see Chapter 4). Nevertheless they are edge-fastened: strakes overlap in a half lap within the thickness of the planking so that, outwardly, they appear to be flush laid; they are fastened together within this half lap by boatbuilders' staples (Greenhill, 1976: fig. 26, no. 5; 1995B: fig. 36, no. 5).

Thus, without other evidence, depictions of hulc planking cannot be interpreted as reverse-clinker planking. Even if it is known, from a caption for example or other textual evidence, that the original planking overlapped, one cannot be sure which form of clinker the craftsman intended to represent since the hulc planking pattern can be achieved using European clinker. (For a worldwide view of hulc planking patterns and reverse-clinker planking, see the Appendix to the present volume.)

Type names

The type names used above for different traditions of boats are those given by the authors quoted. During five seasons of fieldwork in South Asia, we found not only that several different names can be given to one type of boat but also that one particular name may be applied to two (or more) clearly different types of vessel. Deloche (1994: 156) likewise found many inconsistencies in the local terminologies for boats in the Ganges region, while Jansen (1989: 74–5, 236) has noted that, in Bangladesh, 'names change according to use and location, and the same type of boat may have different names in different places'. The scope for confusion, in the minds of people unfamiliar with local usage, is considerable when, for example, *kosha* may mean either 'any boat without a *goloi*' (a specific form of block stem – see page 37) or 'a flat-bottomed boat' (Jansen, 1989: 74–5). However, South Asia is little different in this respect from NW Europe, where boat and ship types (e.g. 'schooner') are seldom clearly defined.

This flexibility of nomenclature, together with the variety of languages in South Asia and the difficulty of transliterating type names, means that, as we shall see below, one cannot be certain that a type of boat named by one author is the same as a boat of the same (or similar-sounding) name noted by another. Only if the two boats concerned had similar hull forms and structure, and similar propulsion and steering arrangements, can we conclude that the two authors were probably describing the same boat type.

Earlier evidence for reverse-clinker boats

Twelfth-century Indian sculptures of boats with reverse-clinker planking, now in Puri, Orissa, in Calcutta and in London, are discussed in Chapter 3.

Seventeenth century

European travellers, merchants and civil servants commented on, and sometimes sketched, South Asian water transport (Hill, 1958). Among these was Thomas Bowrey, a seventeenth-century master mariner and pepper merchant, who wrote an account of his activities in the Bay of Bengal region between 1669 and 1679 that was subsequently published by the Hakluyt Society (Temple, 1905). Bowrey's 'Account' includes brief descriptions and depictions of vessels from the Ganges/Padma region and, of these, two or three may have had clinker planking.

(A) *Putelee*. A large cargo boat.
(B) *Patella/patunga*. A large, flat-bottomed sailing boat of the River Huegli (Temple, 1905: 225, 229, plate 15).
(C) *Purgoo/porka/purga*. An oared, possibly also sailed, vessel which carried cargo between the River Huegli and Pipli and Balasore in Orissa. This type was first mentioned in the Bengal records of the East India Company in 1650 (Temple, 1905: 228, plate 13).

Bowrey did not publish an illustration of the *putelee*, and it may be that (A) and (B) refer to the same boat type.

From the illustrations (only in profile) and the brief descriptions given by Bowrey, it is impossible to tell whether the planking of the original vessels was flush laid ('smooth skin'), reverse clinker or European clinker. This European style of clinker planking was noted in the Ganges/Padma region (Deloche, 1994: fig. 35) by both Hornell (1946: 249–50, plate 29B) and Greenhill (1971: 107–109), so it may have been used there in the seventeenth century and, indeed, earlier.

Late eighteenth century

The Antwerp artist F.B. Solvyns published drawings (with brief notes in the Paris edition, but not in that of Calcutta) of many South Asian boats (Hardgrave, 2001). Three boats from the Bengal/Bangladesh region had, or may have had, reverse-clinker planking.

(A) *Pettoua/petooa* (Solvyns, 1799: vol. 3.9: no. 5) (see Fig. 3.4 below).

A sailing cargo vessel of some size, steered by a median rudder. Solvyns noted that she was 'clinker-built: the boards are upon one another, contrary to our European construction of that kind'. Above the twelve or so reverse-clinker strakes are six other strakes, possibly flush laid, which appear to run to the bow and almost to the stern. Like many traditional boats of South Asia today, this vessel has protruding crossbeams; she may also have had posts. She has a distinct resemblance to the *patia* of twentieth-century Orissa and West Bengal (see Chapter 3).

(B) *Pansway/panswae* (Fig. 2.8; Solvyns, 1799: vol. 3.8: no. 6).

A river boat of the Huegli carrying cargo or passengers. She was propelled by oar or by sail and steered by a rudder on the starboard quarter. The bow is somewhat lower than the stern and there may be block ends (*goloi* – see below). Hulc planking is depicted, but it is impossible to say whether this is reverse clinker. Above these strakes is a single strake which runs near horizontally from end to end.

(C) *Pataily/pataeley* (Fig. 2.9; Deloche, 1994: 161, fig. 21D; Solvyns, 1799: vol. 3.9: no. 7).

A flat-bottomed river cargo boat, with protruding crossbeams, of the Bihar and Benares region. She was propelled by sail and steered by a large-bladed rudder on the starboard quarter. The bow is lower than the stern, but neither seem to have a *goloi*. This vessel has hulc planking at the bow, but not at the stern. The planks were 'laid over each other', which probably means that they were clinker. It is possible, therefore, that this was reverse-clinker planking, but this is far from certain.

Figure 2.8 A *pansway/panswae* of the River Huegli, drawn by Solvyns in the late eighteenth century. Bodleian Library, University of Oxford: FC4(10) Section 8, No. 6.

Figure 2.9 A *pataily/pataeley* of the Bihar and Benares region drawn by Solvyns in the late eighteenth century. Bodleian Library, University of Oxford: FC4(10) Section 9, No. 7.

Early nineteenth century

Lieutenant (later Admiral) F.E. Pâris sketched, painted, compiled measured drawings and commented on many vessels of South Asia. Reverse clinker is not mentioned in his text (1843), but some of his illustrations show hulc planking and two depict vessels which are comparable with Solvyns' (B) and (C) and have similar type names.

(A) *Pansway* (Pâris, 1843: 43, plans 37, 38).

Pâris' illustration shows a vessel with two edge-fastened, near-horizontal strakes rather than the single strake in Solvyns' (B).

(B) *Patile/pataily* (Pâris, 1843: 43, plan 35).

Pâris depicts a boat with edge-fastened hulc planking at both ends, the pattern being modified to match the rising stern. The text identifies the planking as clinker. Although Pâris gives a view of the vessel from the port quarter, it is not possible to decide which way the planking overlaps. Surmounting these clinker strakes are five full-length strakes, reducing to three right aft. The crossbeams protrude. This vessel is propelled by several standing oarsmen, some pushing, some pulling.

Other vessels drawn by Pâris (1843: plans 33 [figs 1, 2], 34 and 43) and also described as *patile/pataily* do not have hulc planking; nor does the *patella/ patunga* illustrated by Bowrey (Temple, 1905: 225, 229, plate 15). It thus seems that the terms *patella/pataily/patile* were used in the seventeenth to nineteenth centuries to describe at least two types of vessel: one with hulc planking and one without.

Solvyns did not state that his *pataily* had reverse clinker, as he did for the *pettoua*. However, both he and Pâris depict hulc planking, and Pâris shows edge-fastened planking, while Solvyns says it was overlapping. It seems likely therefore that they were describing vessels of the same type, but whether with reverse or European clinker is not clear: the balance may lie with the latter.

(C) *A boat from the mouth of the Ganges/Padma* (Pâris, 1843: plan 31, figs 3, 4 and 5).

This boat, propelled by oar and by lugsail, has edge-fastened, hulc planking, with a near-horizontal standing strake and a weather board/washstrake. She is steered by a steering oar over the stern. This was almost certainly clinker planking, possibly reverse clinker.

(D) *Baulca* (Pâris, 1843: plan 32, figs 1 and 2).

This boat is propelled by a lugsail and by oars and steered by a steering oar over the stern. The bow was lower than the stern. She had hulc planking, two near-horizontal strakes and a weather board. Since plank fastenings are not shown, this boat is probably not clinker-built.

(E) *Dinghi* (Pâris, 1843: plan 32, figs 3, 4 and 5).

This boat was propelled by a sprit sail and oars and steered by a steering oar over the (unusual, for South Asia) starboard quarter. The bow is lower than the stern. This boat has edge-fastened, hulc planking, with one near-horizontal strake from end to end. These features suggest that this boat possibly had reverse-clinker planking.

Twentieth century

In several articles (1920, 1924 A and B, 1926, 1930, 1933, 1943 and 1945) and in his major book (1946), James Hornell dealt with many aspects of South Asian water transport. In none of these works did he mention reverse-clinker strakes or describe that pattern of planking which we call hulc planking (Greenhill, 1971: 83–4).

Basil Greenhill undertook fieldwork in Pakistan and Bangladesh between 1950 and 1959, and this has been published in three articles (1957, 1961 and 1966) and in sections of two books (1971; 1995A: 38–46, 254–5). Greenhill divided the Bangladeshi river craft he encountered into six classes, the second of which consisted of round-hulled boats with reverse-clinker planking (Greenhill, 1971: 99–107). These boats were mainly to be found in the northern part of Sylhet District on the headwaters of two of Bangladesh's three great rivers, the Brahmaputra/Jumuna (Jamuna) and the Meghna (Fig. 2.1). Others were seen on the River Pussur, south of Khulna; in the western part of Mymensingh District, near the Brahmaputra/Jumuna; and possibly in Dhaka District near Mograpara Hat.

All the reverse-clinker boats seen by Greenhill had a *goloi* at each end (see Fig. 2.6). *Goloi* is the term applied to the block ends which almost all types of Bangladesh river boat have. Some *goloi* are long and pointed, while others are blunt and rounded. In reverse-clinker/*digekata* boats, *goloi* are broad block-stems, rectangular in plan at their ends and with a wedge-shaped cross-section, which rise from the hull at a shallow angle to the horizontal (Greenhill, 1971: fig. 8). Conversely, on boats without reverse clinker, *goloi* can rise sharply at a steep angle (Greenhill, 1971: 87 [upper], figs 6, 7, 11, 12, 13, 17 and 18). In the expanded logboats extended by reverse-clinker planks that Greenhill saw at Dawki, near the Bangladesh/Assam border, the *goloi* were integral extensions of the boats' logboat bottom. Generally, however, *goloi* are separate timbers, fashioned to the appropriate shape and scarfed to the plank-keel, as seen on some of the large, reverse-clinker boats which, in the 1950s, carried cargo between Dhaka and Sylhet town (Greenhill, 1971: 99–107). Greenhill (pers. comm.) found in the 1950s that some boatbuilders seemed to specialise in the production of *goloi* as a unit with a scarfed section of plank-keel.

Greenhill noted that, in all these reverse-clinker boats, the garboards (first side strakes) were fastened outboard of the logboat base (in extended logboats) or the plank-keel (in plank boats), either in overlapping, European clinker fashion or in a diagonal bevel. The number of reverse-clinker strakes varied from two to three each side in the extended logboats, and from two to six in the planked boats. Above the reverse-clinker strakes there was invariably a standing strake that ran the full length of the boat, each side. A few boats had a weather board above this, which stopped short of the ends.

Extended logboats and plank boats were built plank first, the framing being fitted after the planking was (near) complete. Logboats had no floor timbers, but some had light crossbeams low down. Plank boats occasionally had floors; most had upper crossbeams near sheer level and low crossbeams, sometimes with stanchions to the plank-keel.

Nazibor Rahman, former Secretary General of the Bangladesh Country Boat Owners Association, recently reported (pers. comm.) that, in the 1950s, reverse-clinker cargo boats known as *shoronga* were used in the northern Sundaband, in Kishorganj, Brahmanbaria and Habiganj districts. These boats measured 52 to 62 ft (16 to 19 m) in length, 10 to 18 ft (3 to 5.5 m) in breadth and 4 to 8 ft (1 to 2.5 m) deep, and they had a cargo capacity of 800 to 1500 *maunds* (29 to 56 tonnes). Although Rahman described these boats as 'flat-bottomed', it may be that they were similar to the big cargo boats noted by Greenhill (1971: 87 lower), which had a full form with rounded bilges, i.e. were almost flat-bottomed.

Rahman also noted that, in recent years, reverse-clinker, round-hulled boats known as *Sylheti kosha* carried stone, sand and paddy (rice in the husk) by river from Sylhet to Dhaka, returning to Sylhet with consumer goods. These boats could carry 350 to 400 *maunds* (13 to 15 tonnes) and measured 35/45 x 8/10 x 4/5 ft (10.7/13.7 x 2.4/3 x 1.2/1.5 m). Jansen (1989: 78, fig. 2.12) noted that *Bajitpuri* or *Sylheti patam* boats were also used recently in the stone trade between Sylhet and Dhaka. These boats had about twice the capacity of the *Sylheti kosha* and they appear to have been *binekata* (smooth-skinned) rather than *digekata* (reverse-clinker).

Fieldwork

The aim of the fieldwork undertaken in December 1997 was to follow in Dr Greenhill's footsteps and visit sites in Bangladesh where reverse-clinker boats were built and used, and to document one type fully.

The information given by Greenhill (1971) and by Rahman was used to compile a list of sites where reverse-clinker boats might still be found today. M. Abrar Hossain, a post-graduate student of geography at the University of Rajshahi, undertook an invaluable reconnaissance of these sites over a three-week period in October and November 1997. He visited five sites in Dhaka District, four in Mymensingh and three in Rajshahi, without finding reverse-clinker boats. However, he did find them in Sylhet District: at Bhola Ganj, 15 km NE of Compani Ganj; at Ganeshpur, NE of Chhatak; at Tahirpur, 50 km W of Chhatak; and at Bolla Ghat, on the River Piyain near Jaflong (see Fig. 2.2). In Khulna District he found one reverse-clinker boat at Alaipur but none at six other sites he visited. Hossain had been told in Sylhet that reverse-clinker boats were still used in Barisal District, and in Khulna he was told of similar boats in Kushtia District. These reports could not be confirmed by personal observation.

Sylhet featured prominently in Greenhill's accounts and in Rahman's recollections. Furthermore, Hossain had found reverse-clinker boats on several sites reasonably close together in that region. It was therefore decided to investigate the rivers in a sector WNW to N of Sylhet town and then to move on to Khulna, Kushtia and Barisal. In the event, it proved possible to visit only the sites in Sylhet.

Bolla Ghat

This site is a boat landing place on the River Piyain, a tributary of the River Surma, a few kilometres SSW of a gap which the Piyain has cut through the

Khasi Jaintia mountains along the course of a fault line. Quantities of sand, gravel and stones from the mountains to the north are carried south by this river until deposited further downstream as the river slows on entering the Sylhet plain. This deposition is greatest during the monsoon season, when severe flooding caused by flash floods forces rivers to change course. Severe flood waters in 1988 caused the River Piyain to cut a second channel some kilometres to the southeast of its original bed, leaving an island between the two courses. The new cut, known locally as the Jaflong stream, is now the main channel and Bolla Ghat lies on it.

During the dry season, stone and its derivatives are 'quarried' from this river bed and its surroundings. Hundreds of boats are used, and thousands of people employed, in the extraction and transport of this material.

Bhola Ganj

Downstream of Bolla Ghat, at Compani Ganj, the River Piyain is joined by the River Dholai, which rises in the Khasi Jaintia mountains to the north. Bhola Ganj lies some 9 km up the Dholai, which in the wet season is very broad but in the dry season very shallow. Boats are built at Bhola Ganj to transport stone and its derivatives, mined further upstream, down to Chhatak where the rivers Piyain and Bagra join the Surma.

Ganeshpur

Ganeshpur is on the lower, meandering course of the River Piyain, to the northeast of Chhatak. Generally, this river is sufficiently deep for boats to be used here throughout the year. Boats for the transport of stone are built at Ganeshpur.

Choice of site

At these three sites there were many reverse-clinker boats, most of them involved in the extraction and transport of stone and its derivatives (Fig. 2.10).

Fieldwork was concentrated on Bolla Ghat where the largest number and the largest variety of reverse-clinker boats were found. Most of these were 'country boats' (that is, they were wooden boats built by traditional methods and propelled solely by pole, paddle, oar or sail), but there were twenty or so with engines. One beached boat (Boat B) was recorded in detail; two others (Boat C and Boat CAB) were partly documented; and notes were compiled on others, including Boat D and Boat AB. Boat A was being repaired by a *mistri* (boatbuilder) on the dry river bed, and many of the timbers, including the *goloi*, were dismantled and replaced by new ones. Detailed notes were made on this work, and the *mistri* demonstrated some of his methods. Except for AB,

Figure 2.10 Reverse-clinker boats engaged in the extraction and transport of stone and its derivatives on the River Piyain at Bolla Ghat, Sylhet District, December 1997. Graded stockpiles of boulders, stones and shingle are in the foreground.

which was motorised, all these boats were traditional country boats of one type or another.

Other boatbuilders were interviewed at Ganeshpur and at Bhola Ganj, so that any variations in boatbuilding techniques could be recorded. It proved impossible to visit Tahirpur (some 50 km west of Chhatak), but we were able to draw on the notes taken by M. Abrar Hossain during his reconnaissance.

Types of reverse-clinker boat

Three main types of reverse-clinker boat were identified at Bolla Ghat: they are mainly distinguishable by reference to their cross-section and their slenderness (length-to-beam) ratio L/B (see Annex I in Chapter 9). Two of these types have a motorised variant. Type 1 (Fig. 2.11) is generally similar to the boats seen by Greenhill in this region in the 1950s (Fig. 2.12). Type 2 (Fig. 2.13) appears to be a simplified version of Type 1. While Type 3 (Fig. 2.14) is a related, but distinctively shaped, boat possibly developed in recent times.

At Bolla Ghat, Type 1 and Type 2 boats were generally called the Bengali equivalent of 'large boat' or 'little boat', depending on size. Since the Type 1 boats generally seemed to correspond to the Sylhet round hull, reverse-clinker boats (see Fig. 2.12) described by Greenhill (1957: 123–8; 2000: fig. 8;

Figure 2.11 A Type 1 boat (Boat A) under repair on a dried-out part of the river at Bolla Ghat. The boat has one garboard strake and five reverse-clinker strakes, laid hulc fashion each side and capped by a full-length reverse-clinker strake. Reverse-clinker washstrakes have been removed from the top of the sides.

1971: 99–110; 1995A: fig. 327) nearly fifty years ago, we decided to give them the codename *Sylheti nauka*. Occasionally we were told that Type 2 boats were known as *kosha*, but this is not a precise term. We therefore called them *barki*: this was the name given by Rahman to a current type of Sylhet reverse-clinker boat which appears to have similar features to our Type 2 boats. The Type 3 boats were widely known in Sylhet as *Chhataki nauka* (i.e.'boats from Chhatak'), and this is the name we use.

Type 1: the Sylheti nauka

This Type 1 reverse-clinker boat (see Fig. 2.11) has an elongated, double-ended 'boat shape' in plan, a stern that is higher and steeper than the bow,

Figure 2.12 A reverse-clinker boat under construction in Sylhet District in the 1950s. This boat has some similarities with the Type 1 boat in Fig. 2.11. Copyright: Basil Greenhill.

and a round bottom with flaring sides in section. We were able to examine in detail only one boat of this type, Boat A, which was undergoing a major overhaul close to the river, but several others were noted in use. Boat A measured 9.33 x 1.29 x 0.40 m with the shape ratios: L/B (length to beam): 7.2; B/D (breadth to depth): 3.2; L/D (length to depth): 23.3 (see Annex I in Chapter 9).

Type 1 boats have a plank-keel with a fully shaped *goloi* at each end. The plank-keel, consisting of at least three planks laid end to end, rises gently upwards towards its ends to meet the curving *goloi*. The garboards (first side strakes) are fastened to the plank-keel in a bevelled lap. The remaining strakes are laid hulc fashion, with five or more reverse-clinker strakes, each one shorter than the one below. Above these strakes there is another reverse-clinker strake which runs almost the length of the boat, from *goloi* to *goloi*. There is also a final reverse-clinker strake which stops short of the ends: 'weatherboard' or 'washstrake' is not quite the right term for this, but there does not seem to be a better alternative.

The framing of Boat A consists of floors and crossbeams. There are four floor timbers running across the bottom from the seams between the first and second side strakes. These floors have limber holes centrally and are joggled to fit against the lowest side strake. Six crossbeams are fitted at the level of the upper full-length strake.

Figure 2.13 A Type 2 boat (Boat B) being recorded at Bolla Ghat. The bow is in the
foreground. Two truncated reverse-clinker strakes each side are capped by a
standing strake with a reverse-clinker washstrake above it.

Type 2: the barki

These boats (see Fig. 2.13) have a flat bottom, consisting of a central bottom
plank ('foundation plank') and two outer bottom planks, otherwise they are
similar in shape and size to the Type 1 boats. Such boats measure *c.* 9–10 x
1.2–1.5 x 0.40–0.45 m. Their shape ratios are: L/B *c.* 7; B/D *c.* 3; L/D *c.* 20.
The *barki* has vestigial hulc planking, with truncated reverse-clinker strakes
which are fewer and broader than the strakes in the *Sylheti nauka*. Above the
reverse-clinker strakes there is a standing strake (not overlapping) running
full length from block-*goloi* to block-*goloi*; and above this there may be a

43

second full-length standing strake or a reverse-clinker strake which stops short of the ends.

Like the *Sylheti nauka*, the *barki* has upper crossbeams at standing strake level – in this case two to four. Instead of floors, however, the *barki* has twenty or so lower crossbeams at the level of the seam between the lowest and second side strakes; some of these are supported on short pillars.

Barki boats were seen not only at Bolla Ghat but also at Chhatak, where they were used as ferries and as cargo carriers, and at other places on the tributaries of the River Surma.

The motorised barki (Type 2.1)

Motorised boats used as ferries at Bolla Ghat are somewhat greater in size but similar in shape to the *barki*, and their planking patterns are similar. The principal differences are that the timbers of the powered boats are generally of greater scantlings (i.e. dimensions) and their framing is reinforced to cope with the installation of an inboard engine. There are side timbers in between the floors; the crossbeams rest on beam shelves; and inwales are fitted along the top strakes to stiffen the hull. Bottom boards are also fitted, and there is a special structure for the port-side rudder. Some of these ferries have a short awning over the engine, while others have a longer one to protect passengers.

Type 3: the Chhataki nauka

In plan, these Type 3 boats (see Fig. 2.14) are 'logboat-shaped', being relatively much longer than Types 1 and 2 in relation to both breadth and depth. Like Type 2 boats they have a flat bottom but, unlike boats of Types 1 and 2, they are near-rectangular in section. Such boats measure *c.* 11 x 1 x 0.30 m; their shape ratios are L/B *c.* 11; B/D *c.* 4; L/D *c.* 40.

These boats have a plank-*goloi*, rather than a block-*goloi*, at each end; their bottom planking is similar to that of the *barki*, but each plank is somewhat broader; and they have vestigial hulc side planking with two or three reverse-clinker strakes and one full-length standing strake. No additional standing strakes or washstrakes were seen on any *Chhataki nauka*. Towards each end there is a crossbeam at standing strake level, otherwise the only framing timbers are short 'floors' extending merely the breadth of the central bottom plank. These have more the appearance of foot-grips than framing; however there is the possibility that they are skeuomorphs of transverse timbers used to keep open the sides of expanded logboats which, in earlier times, formed the base of some Sylhet reverse-clinker plank boats (Greenhill, 1971: 100–101; 1995A: 107).

Chhataki nauka were seen not only at Bolla Ghat, where they were the predominant boat type, but also at Bhola Ganj and Ganeshpur, and at many other places on the headwaters of the River Surma.

Figure 2.14 Two Type 3 boats on the River Piyain at Bolla Ghat. The two truncated reverse-clinker strakes each side are capped by a standing strake. Note the near-rectangular transverse section.

The motorised Chhataki nauka (Type 3.1)

The motorised version of the *Chhataki nauka* boat is generally stronger all round, has more framing and is somewhat greater in size. Their shape, especially the cross-section, is generally similar to that of the Type 3 boat, but they are broader in relation to length and much deeper in relation to both length and breadth, their shape ratios being: L/B *c.* 8; B/D *c.* 2; and L/D *c.* 16.

They have plank-*goloi* and their bottom planking is similar to that of the *Chhataki nauka*, but broader. Their side planking is similar, but above several strakes of reverse clinker they have one or more full-length strakes which overlap in either reverse-clinker or European clinker fashion. The top strake may have a rubbing strake outboard.

The framing of these boats consists of: floor timbers (rather than short 'floors') extending to the seam between the first and second side strakes; side frames, to the upper ends of which an inwale is fastened; and an intermittent beam shelf supporting crossbeams at the level of the upper full-length strake.

A khashia *boat*

One specific small boat (CAB) was said to be a *khashia* boat, probably because it was owned by a group of people known as 'Khasi' (Greenhill, 1971: 100): she appeared to be a hybrid Type 2/Type 3 boat (Fig. 2.15). She measured 6 x 1.01 x 0.33 m, with shape ratios of L/B: 5.9, B/D: 3.1 and L/D:

Figure 2.15 A *khasi* or *khashia* boat (CAB) at Bolla Ghat being recorded. Stern in foreground.

46

18, and thus had a shape similar to Type 2. On the other hand, although she had four rather than two crossbeams, she had a Type 3 plank-*goloi* and a series of short 'floors' which extended only across the plank-keel, comparable with those fitted in Type 3 boats. See also the boat noted in Sylhet (Fig. 2.16) during the 1950s by Greenhill (1995A: 107, fig. 105).

Documenting the boats

The aim of our research at Bolla Ghat was to record one of the traditional reverse-clinker boats in detail and to note the significant characteristics of other boats. There were so many boats in sight, both up and downstream of

Figure 2.16 A small boat photographed in the 1950s in Sylhet District. This boat has similarities with the *khasi* in Fig. 2.15. Copyright: Basil Greenhill.

the main landing place, that at first sight the task of identifying a typical boat seemed daunting. Most of them were in use, however, and seemed set to remain so during the daylight hours of every day. The search was therefore soon narrowed down to a particular place on the NW bank of the River Piyain where there was a group of four beached country boats, with others, including motorised ones, moored in the river only a few yards away. One of the four beached boats (Boat A) was being repaired: this was a *Sylheti nauka*. Boat D, a *Chhataki nauka*, was upside down being tarred. Unattended boats B and C were *barki*: it was decided to record the smaller one (B) in detail – the overall measurements of this boat were 8.57 x 1.16 x 0.44 m.

Observation of the repair work on Boat A proved of equal value to the compilation of the measured drawing of Boat B, since, during our five days on site, both ends of A were dismantled and partly rebuilt with new timber. Since Boat B was covered in tar, as were all the other boats, details of its structure were sometimes obscured. For example, the overlap of the reverse-clinker planks could only be detected by the most careful physical examination. On the other hand, when the ends of Boat A were dismantled structural details became much clearer; and when the builder replaced planking he kindly demonstrated the nailing technique.

After recording Boat B and studying in detail the work in progress on Boat A, we made notes on boats C and D and a cross-section of C was measured. Beached further inland, but still within the flood-time river bed, was a motorised version of the *Chhataki nauka* (Boat AB): the main features of this boat were also noted. Subsequently our interpreter, M.A. Hossain, compiled under supervision an outline measured drawing of the small *khashia* boat (CAB) which had been beached near the group of four boats.

Boatbuilding

Materials

At Ganeshpur, species of wood known as *jam* and *mango* (*Mangifera* sp.) are used for the bottom planking and *gumma* (a very red coloured timber) for the side strakes. In the early part of the twentieth century, Bangladesh boatbuilders used *sal* (*Shorea robusta*) and teak (*Tectona grandis*), but these timbers became difficult to obtain and such timbers as *sundari*, *garjari*, *zarail*, *kathal* (jackfruit), *shilkoroi* and *koroi* are now used (Jansen *et al.*, 1989: 9, 233).

Building a Sylheti nauka (Type 1)

We were unable to observe the building of a Type 1 boat, and so the building sequence and the woodworking techniques described below are based on detailed observations of intermittent repair work on Boat A (Fig. 2.17) and on videotapes of the boatbuilder working.

Figure 2.17 The stern of Boat A (Type 1), under repair at Bolla Ghat. To the left of the plank-keel there are a garboard and three reverse-clinker strakes, each one shorter than the one below. The ends of these strakes (and two others out of the picture) are capped by another reverse-clinker strake running from *goloi* to *goloi*. The 'ghost' of a short washstrake (also in reverse clinker) can be seen along the upper edge of the full-length strake.

It is particularly to be regretted that we did not see the early stages of the building process, when it might have proved possible to deduce something about how such boats were 'designed'. Since these reverse-clinker boats are all built plank first, much of that 'design' process was probably 'by eye': that is, the builder fashioned the planking as he thought best to give the shape of hull he had in his mind's eye. The fact that the *mistri* repairing Boat A appeared to use his eye rather than measurements, when shaping new ends for the boat, gives support to this hypothesis.

The plank-keel and the goloi

For its major refit, Boat A was positioned on an informal group of three heaped mounds of beach sand; however, a more rigid arrangement was probably needed when she was originally built. Building a *Sylheti nauka* therefore probably begins with the laying of the plank-keel on some form of simple stocks (see page 56). Boat A's plank-keel was seen to be fashioned from part of a half-log with the pith uppermost; it thus has an almost natural bevel along its edges and a slightly curved profile underneath. Near each end of the plank-keel a drain hole is cut and filled with a removable wooden bung.

The lower end of the *goloi* has a vertical face, to the underside of which the plank-keel is scarfed. From this face towards the extremities of the boat, *goloi* are shaped as a block stem which slowly curves upwards. The after *goloi* rises somewhat more steeply, and to a slightly greater height, than the forward one.

Shaping and fastening the lowest side strakes

The first side strakes are, like the plank-keel, *c.* 30 mm thick and they extend as far as the shoulders of the *goloi*. They are set in a bevel, at an angle to the plank-keel, thus beginning the transverse curve of the hull. Towards the ends of the boat these planks begin to rise to match the rising element of the *goloi*, and they taper in breadth.

After being worked to shape, the inner edges of the first side strakes are given a bevel and tarred. A series of recesses for iron staples is then chiselled in the inner face of the plank-keel near its edges, the spacing being *c.* 48 to 60 mm. The two strakes are then positioned in succession alongside the plank-keel, a caulking of jute is inserted in the seams, and the planks are brought firmly together and held in position with the aid of wooden clamps. The strakes and plank-keel are then fastened together inboard within the bevelled lap, using iron staples cut from thin plate; these measure *c.* 66 x 6 mm and are pointed at one end. The pointed end is driven into the inner face of the first strake, near to its inboard edge (Fig. 2.18). The protruding part of the staple is then hammered around a metal rod towards the recess cut in the plank-keel (Fig. 2.19); the rod is removed, and the hook thus formed is driven into the recess until it lies flush. This process is repeated along both edges of the plank-keel.

The keel and strakes are then fastened together outboard by a similar process, but this time the hooked staples are driven into a recess in the outer face of the first strakes (Fig. 2.18). The ends of these strakes are fastened inboard and outboard to the shoulders of the *goloi* in a similar manner.

Fastening the reverse-clinker strakes

The reverse-clinker strakes are each shaped and bevelled to give the hull shape needed; precisely how this is achieved is a major 'mystery' of the

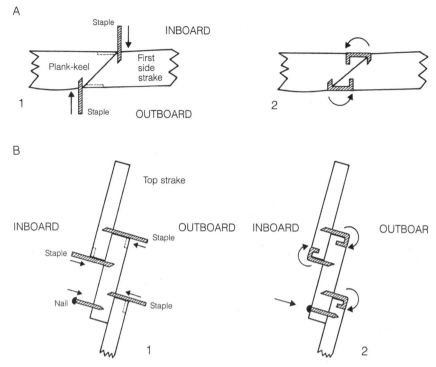

Figure 2.18 Diagram to show the method of fastening the planking of a *Sylheti nauka*: A. Fastening the garboards to the plank-keel; B. Fastening together the reverse-clinker strakes.

boatwright's art. The longer strakes consist of more than one plank; they are scarfed together in vertical laps which vary in length from 100 to 190 mm. There is no obvious pattern to the direction in which these scarfs open.

The *second* side strake (i.e. the first reverse-clinker strake) runs from one *goloi* to the other. Each succeeding reverse-clinker strake is shorter than its predecessor, ending well before that station where it could no longer form a reverse-clinker lap with the plank below. This arrangement can be clearly seen in Figures 2.11 and 2.17. These strakes gradually narrow towards bow and stern, and they each lie in a gentle longitudinal curve which flows upwards from the centre of the boat towards each *goloi*. The last of these short reverse-clinker strakes is shaped to fill the gap in the curve made by the ends of these strakes.

Each reverse-clinker strake overlaps the strake below by *c.* 25 mm. They are fastened to each other by both nails and staples. After an individual strake has been fashioned to shape, its lower outboard face is tarred, caulking is applied, and the strake is then positioned lapping the one below. The two strakes are first fastened together by driving iron nails, used as spikes,

51

Figure 2.19 Fastening a strake to the plank-keel of Boat A. The squared end of the staple is bent around a metal rod and then hooked by driving it into the recess cut in the plank-keel.

through this lap from inboard: these nails enter but do not penetrate through the lower strake. Staples are then used to fasten the planks together from outboard. The pointed end of each staple is driven into the outer face of the upper strake, alongside the upper edge of the lower strake. The protruding length of the staple is then hammered around a metal rod and the hook so formed is driven into a recess which has been worked in the outer face of the lower strake near its upper edge (see Fig. 2.18).

Fastening the top strakes

The upper reverse-clinker strake which runs from *goloi* to *goloi* laps the upper edge of the highest short reverse-clinker strake and laps the ends of the other strakes. The protruding ends of the lower strakes are then chamfered to give a smooth curve to the outer hull. At its ends this upper strake stands on, and is fastened to, the upper surface of the *goloi*. Staples, both inboard and outboard, are used to fasten this strake in position (see Fig. 2.18). The washstrake, which ends *c.* 0.50 m from each *goloi*, is fastened in a reverse-clinker lap in a similar manner.

The framing

The floors nailed to the lower planking of Boat A were relatively substantial, being *c.* 40 mm broad (siding) and *c.* 75 mm deep (moulded) on the boat's middle line. A single limber hole, for the free passage of bilge water, had been cut through them centrally above the plank-keel. The floors were notched (joggled in the 'reverse' manner) on their outer face so that they fitted closely against both first and second strakes on both sides of the boat.

The six crossbeams in Boat A were fastened by spikes driven from outboard through the top strake and into the ends of the beams.

Building the barki (Type 2)

Since the *barki* is flat-bottomed, it could in theory be built on any level surface. However, at some stage, access underneath is required so that fastenings may be driven into the bottom from underneath. This could be achieved by turning over the bottom of these small boats. On the other hand, the flat-bottomed *Chhataki nauka* is built on simple stocks, and it seems likely that the *barki* is similarly built.

The *barki* (Fig. 2.20) differs from the *Sylheti nauka* in four main ways:

i)   It has a flat bottom consisting of a central bottom plank and two outer bottom planks, rather than a rounded hull formed by a plank-keel and two garboard strakes.

ii) It has two to four relatively broad reverse-clinker side strakes and a standing strake, rather than six and more narrower ones; the lower strakes are truncated with vertical ends, rather than the horizontal ends of the *Sylheti nauka*.

iii) It has many lower crossbeams, rather than a few floor timbers.

iv) It has an intermediate form of *goloi* which does not have a vertical face at its lower end, as do the *goloi* in the *Sylheti nauka*, but tapers in thickness (moulding) from around the waterline downwards to their scarf with the plank-keel. The breadths (sidings) remain fairly constant from the plank-keel upwards until the 'shoulders', where they increase substantially.

Otherwise, the two boat types have a comparable shape and structure and, in general, are built in a similar manner.

The bottom planking

The midships element of Boat B's central bottom plank is 2.36 m long, 160 mm broad (sided) and 30–40 mm thick (moulded). Further planks are scarfed to each end of this plank and they in turn are each scarfed to a *goloi* in a horizontal scarf: these scarves both open aft. A central longitudinal groove is worked along the upper surface of the plank-keel and the two *goloi*, so that water runs to a position from which it can be bailed.

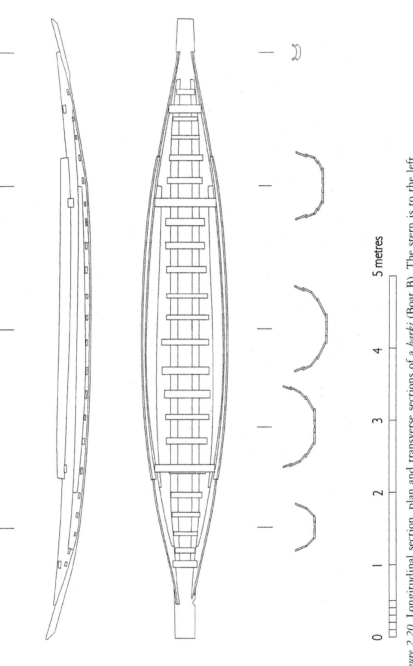

Figure 2.20 Longitudinal section, plan and transverse sections of a *barki* (Boat B). The stern is to the left.

0   1   2   3   4   5 metres

Boat B's outer bottom planks are 170–180 mm broad at the midship station and retain this breadth for much of their length. They are fastened to the central plank in a similar way to the garboard/plank-keel of the *Sylheti nauka* but remain in the horizontal plane until near the ends of the boat, when they are angled inwards and upwards to conform to the rising *goloi* and the reduction in breadth of the boat.

The side planking

Boat B's first reverse-clinker strake is 25–30 mm thick and 160–180 mm broad near amidships. It is bevelled on its inner lower edge, where it meets the outer bottom plank in a reverse-clinker lap, and chamfered on its upper, outer edge at the ends to conform to the hull curvature there.

Boat B had only one other reverse-clinker strake (Fig. 2.20). Boat C, on the other hand, not only had more such strakes, but it had four to port and only three to starboard (Fig. 2.21). The *barki* strake fastenings are similar to those of the *Sylheti nauka*, spacings between fastenings varying from *c.* 48 to 60 mm.

All the *barki* observed had, above their reverse-clinker strakes, a standing strake running from the upper face of one *goloi* to the upper face of the other. Boat B's standing strake was *c.* 200 mm broad amidships, narrowing to *c.* 40 mm towards the ends to allow for the rising *goloi*. Such strakes are fitted flush with the lower strakes and do not lap them. Staples, both inboard and outboard, are used to fasten this standing strake to the *goloi*, to the upper edge of the highest reverse-clinker strake and to the ends of the other strakes. The standing strakes thereby 'seal off' the clinker strakes and form the upper edge of the boat's sides.

Some *barki* (for example, Boat B) had a washstrake above the standing strake; others (for example, Boat C) did not. Such boards are fastened in

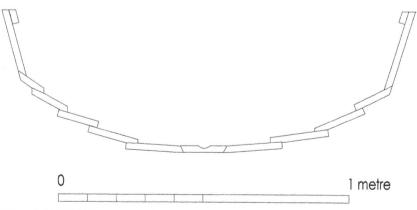

0   1 metre

Figure 2.21 Midships section of Boat C.

reverse-clinker fashion, overlapping the standing strake by approximately 50 mm, but do not extend as far as the *goloi*. Boat B's washstrake was between 120 and 150 mm in breadth and *c.* 20 mm thick, and was fastened to the standing strake in a similar way to the fastening of the reverse-clinker strakes. Boats without a washstrake may have a narrow inwale along the upper edge of the standing strake.

The framing

Boat B had 22 lower crossbeams at intervals of 220 to 300 mm. They were 20 to 50 mm thick (moulded), 50 to 110 mm broad (sided), and were positioned 40 to 60 mm above the bottom planking, at the level of the first reverse-clinker strake or, towards the ends, at the upper edge of the outer bottom plank. These beams were fastened to the strakes by nails driven from above through the beam and into the planking; some of them were supported underneath by a short pillar (for example, lower crossbeams 9 and 10 from the stern in Fig. 2.20).

Boat B also has four upper crossbeams, two towards each end of the boat, at standing strake level. These beams are between 100 and 120 mm broad and 40 to 80 mm thick. They were fastened by two nails driven through the planking from outboard into each end of the beam.

Building the Chhataki nauka (Type 3)

We visited two sites, Ganeshpur and Bhola Ganj, where *Chhataki nauka* were being built on stocks in an open-sided shed with the planking held in position by struts from both floor and roof. We were unable to observe the early and late building phases, however we did attempt to discuss the 'design' of such boats with the *mistri* (builder).

Aspects of design

We were told that all such boats are roughly the same size: 26 cubits in length and 2 cubits in breadth – *c.* 11.77 m by 0.91 m. The main builder's aid seemed to be a measuring stick 44 inches (1.11 m) in length and marked with an unnumbered scale (Fig. 2.22). The *mistri* evidently used this stick to check symmetry from an overhead centreline; he possibly also used it to position the 'floors' on the boat's central bottom plank. A second 'design' aid appeared to be a set of six bamboo sticks of nearly equal lengths (they ranged from 0.265 to 0.29 m). These were used, in a way that was not entirely clear, to mark out the breadth of the central bottom plank at various stations. A chalk line was then used to delineate the complete shape, and the plank cut to size. We understood that a similar system was used to derive the shape of other planks.

Figure 2.22 Tools used by boatbuilders at Bhola Ganj to build Type 3 boats. The building aids discussed in the text are in the foreground with the photographic scales.

Main features (Fig. 2.23; see also Fig. 2.14)

There are only eleven main structural elements of the *Chhataki nauka* and they appear to be standardised in shape. They are: two *plank-goloi*, a centre bottom plank, two outer bottom planks, two pairs of reverse-clinker strakes and a pair of standing strakes. This boat is thus generally similar to the *barki* but differs from that type of boat in the following ways:

i) The *Chhataki nauka* has a much greater L/B ratio.
ii) It has plank-*goloi*, rather than block-*goloi*, and they are at a slightly steeper angle.
iii) The first reverse-clinker strake is fashioned and positioned so that the boat's cross-section is more rectangular than that of the *barki*; the second reverse-clinker strake and the standing strake continue this process.
iv) Although the *Chhataki nauka* has upper crossbeams like the *barki*, these are only near the ends; instead of lower crossbeams it has short 'floors' fastened to the central bottom plank.

The elements of the *Chhataki nauka* were fastened together in a similar manner to the *barki*. The only technique not previously met on Bangladesh reverse-clinker boats was that a dung mixture was smeared on the plank seams as a waterproofing.

57

Figure 2.23 A *Chhataki nauka* (Type 3 boat), with a load of sand, being propelled by oar and pole.

Production rate

It was clear from conversation with the owner of the boatyard at Ganeshpur that there was great demand for the *Chhataki nauka*. During the six months of the year in which boatbuilding is undertaken, 120 such boats were built in his yard alone. We understood that one boat could be built in less than two days. Because of the crowd which assembled as we approached the site (as is often the case), it was not possible to estimate the number of people actually needed to build a boat.

Boat operations

Uses

Apart from a small region in the northwest of the country, Sylhet District is the only source of stone in Bangladesh. Although the great majority of buildings are built of brick, stone and its derivatives have long been used for foundations and for cement. Once the fast-flowing streams from the Himalayas enter the broad plains of Sylhet, waterborne materials are deposited. During the dry season from November to May, when rivers are narrower, shallower and slower, these materials have been extracted for as long as anyone can remember (Fig. 2.24) and transported by boat.

Figure 2.24 Quarrying in the dried-out bed of the River Piyain at Bolla Ghat, Sylhet District.

The Sylheti nauka *and the* barki

Some boats of these two types were involved in the stone industry, but several were laid up, beached or under repair. Others were occasionally seen being used as 'household' boats, as general-purpose ferries along, as well as across, rivers and as market boats taking produce to and from permanent (*ganj*) and temporary (*hat*) markets.

The Chhataki nauka

The demand for stone markedly increased some twelve years ago when the road and bridge building programme was expanded. It seems possible that, at about that time, the traditional design of small river boat in the River Surma headwaters was modified to produce the *Chhataki nauka*, a boat of simple, standardised form specifically designed to carry high-density loads such as stone and sand, and one that is more quickly, and therefore more cheaply, built.

The several hundreds of *Chhataki nauka* seen during our fieldwork in Sylhet were involved in the extraction and transport of stone and its derivatives. At Bolla Ghat (see Figs 2.10 and 2.24), for example, where a number of different extractive methods are used, these boats were a base for dredging the river bed for stone, using buckets flung on a line; while in the deeper parts free divers plunged with a bucket from a dive boat into depths down to 20 ft (6 m). Larger stones and boulders were quarried by hand from the beds of now-dry parts of the river, and from dried-out former river beds, as water in these 'quarries' was pumped out. *Chhataki nauka* were used to ferry the recovered materials back across the river.

The stone recovered in these ways is sorted and stored at convenient sites on the river bank in five different grades: sand, gravel, shingle, stone and boulders (Fig. 2.10). At Jaflong, where there is now a crushing plant, some of this material is turned into cement. Since large boats cannot use the river near Bolla Ghat during the dry season, lorries in increasing numbers transport much of this stockpiled material direct to the sites where it is needed; this presupposes that any damage that roads and bridges have sustained during the wet season has been repaired.

Where the headwater rivers remain navigable in the dry season, as at Ganeshpur and at Bhola Ganj, the several grades of material are loaded into *Chhataki nauka* and taken downstream to Chhatak. These boats occasionally go singly, propelled by paddle or pole, but where the river is sufficiently deep they are paired, side by side, and towed in a train by a powered boat.

Motorised boats

At Bolla Ghat in the dry season the large, motorised versions of the *Chhataki nauki* were inactive, since the river levels were generally too low for them to carry an economic load. Some of the strengthened and motorised variants of the *barki*, on the other hand, were in use as ferries, but propelled solely by pole or paddle. In the deeper rivers these *barki* acted as tugs, using their engine power to tow downstream trains of paired *Chhataki nauka* deeply laden with stone. In the wet season, it is understood that the motorised *Chhataki nauka* are brought into service to move materials downstream to Chhatak.

Larger craft

Chhatak, where there has been a cement factory for several years (Jansen *et al.*, 1989: 142, 244), is the furthest point upstream that large craft from Dhaka now venture; these vessels transport stone and its derivatives, including cement, downstream. Nazibor Rahman has noted that traditional country boats (i.e. those without engines) are still in use, though fewer in number, on this long-distance route, Chhatak to Dhaka. These boats include a type of *kosha* which has reverse-clinker planking. We were unable to identify any *kosha*, but Rahman described them as round-hulled craft fitted with an awning so that selected cargo could be kept dry. They measure 35–45 x 8–10 x 4–5 ft (11–14 x 2.5–3 x 1–1.5 m) and carry 350 to 400 *maunds* (13 to 15 tonnes) of cargo. In addition to stone derivatives, these boats carry Sylhet paddy to Dhaka and return with consumer goods. It seems likely that this was the type of large boat noted by Greenhill in the 1950s

(Fig. 2.6) which carried 'a cargo of cement dry all the way from Sylhet to Dacca' (Greenhill, 1971: 102–104, fig. 87 lower).

The future of country boats

In the 1980s country boats still predominated in the transport of stone and sand (Jansen *et al.*, 1989: 136–8, 244). However, the dredging of the River Surma as far as Chhatak soon resulted in increasing competition from motorised vessels (Jansen and Bolstad, 1992: 21). Although powered boats, supplemented by lorries, may in the future drive more and more traditional boats out of long-distance haulage, it seems likely that the basic *Chhataki nauka* will remain irreplaceable in those headwater rivers where the sand and stone is extracted as long as the demand for these materials persists, since these sites are difficult of access by road and because these waters are generally too shallow for larger boats.

The traditional *Sylheti nauka* and the *barki* are much less specialised. Some may continue to be used in the stone industry, but their main role will probably be, as now, that of 'household' boat, to and from markets, as ferries, and for general socialising, in a region where no place is more than a few kilometres from a waterway. Their hull form makes them readily usable in these tasks in most river states. Although some were seen at relatively deep draughts, none had as little freeboard as that at which the specialised *Chhataki nauka* was regularly operated (see Fig. 2.23). It seems likely that the *barki* will continue to be built in small numbers. The future of the traditional *Sylheti nauka* is uncertain: more skill appears to be needed to build one than is needed for the *barki*, and they are now rare. The type of boat that Greenhill saw in the 1950s may soon no longer exist.

Where the balance between motorised and traditional country boats will be struck is still uncertain. Increasing natural siltation, the extraction of water and the construction of embankments in the delta all tend to decrease the effective depth of the rivers, while dredging increases it. In general, the deeper the rivers, especially the River Surma, the more motorised boats will be used for long-distance haulage (Jansen *et al.*, 1989: 252–3). The Government holds the key to equilibrium: it can either continue to finance large-scale river dredging, training and embankment, or it can support the country boat users. In fact this is only one aspect of a fundamental Bangladesh problem: how to prevent the entire river system from falling into disuse, a situation which would lead to environmental, economic and social disaster.

The boat's crew

Chhataki nauka, *barki* and *Sylheti nauka* are generally crewed by one man, occasionally two; indeed one man can handle two *Chhataki nauka* when they are paired side by side. The strengthened *barki* used as ferries have two men.

We were unable to determine the crew either of a *Chhataki nauka* under sail or of the large, motorised version of this type of boat, since we never saw such boats in use.

Propulsion

All three types of country boat can be poled, paddled or rowed (Fig. 2.23), or be towed by another boat, while some *Chhataki nauka* can be sailed. The pole is generally used by a man standing on the after *goloi*, the single-bladed paddle by a man sitting either on the after *goloi* or on an after upper crossbeam.

Most *barki* and *Sylheti nauka* have a simple oar pivot of bamboo fastened to the top strake on one quarter, forward of the *goloi* and convenient to an upper crossbeam which can be used as a thwart. The oar was seen used in the sit/pull mode.

At Bhola Ganj some *Chhataki nauka* had a mast beam above a simple mast step timber fastened to the bottom of the boat: sail was said to be used on these boats in the wet season. During December 1997 we saw sails used only once: single sprit sails on *binekata* boats transporting sand on a river close to Sylhet Airport. Since the mast steps seen on the *Chhataki nauka* were about one third the waterline length from the bow, it seems likely that a sprit sail was also used in these boats. In such narrow boats, use would be limited to light, fair winds.

No fittings for sail were seen on any of the *barki* boats we examined. However, simple forms of mast and sail need very little in the way of permanent fittings. In the Orissan *patia* (see Chapter 3), for example, there is no mast step and the mast stands directly on the bottom of the boat (Blue *et al.*, 1997: 202). It is therefore possible that some *barki* boats are, or have recently been, propelled by sail.

Steering

The country boats we observed were steered by the same means as they were propelled; when sail is used, steering is by pole or paddle. The motorised boats, on the other hand, have a rudder on the port quarter which is pivoted on the port side of the *goloi* and on a gallows-like wooden structure a little further forward so that the rudder lies at *c.* 50 degrees to the horizontal (Fig. 2.25). To steer or to alter course, the rudder is rotated around its own longitudinal axis by a tiller. There appears to be enough slack in the pivoting system, however, to allow the steersman to manoeuvre his boat, when needed, by using short sculling (side-to-side) strokes with the rudder.

Mooring

When not in regular use these boats are generally berthed clear of the water on the sloping river bank. At Bolla Ghat, where a few wooden landing stages

Figure 2.25 A motorised *barki* with a rudder pivoted on the port side of the after *goloi* and on a gallows-like structure further forward.

have been built, some boats are moored by a light line direct to the landing stage or alongside another boat. The smallest reverse-clinker boat seen at Bolla Ghat (Boat CAB) had a wooden fairlead on the forward *goloi* which could have been used for mooring or for a tow rope. Out of season, or otherwise when not needed for a considerable time, *Chhataki nauka* are sunk in shallow water (see Fig. 2.14) to minimise drying out and the consequent distortion and splitting. Most of the *barki* boats examined had a plug hole near each end, filled by a wooden bung, so these boats too could have been similarly sunk, although this practice was never observed during our fieldwork.

Discussion

Fieldwork in Sylhet District has led to the hypothesis that three main types of traditional river boat with reverse-clinker planking are built there today. These are the types we have called the *Sylheti nauka*, the *barki* and the *Chhataki nauka*. There remain at least two other types to be investigated: the larger vessels that may still be used on the River Meghna below Chhatak; and the boats used around Khulna.

In this connection it is of some interest to note that there is a boat model (Fig. 2.26) in the Science Museum, London (1929–1102: now on loan to the Birla Museum, Calcutta), which resembles the Khulna boats seen by

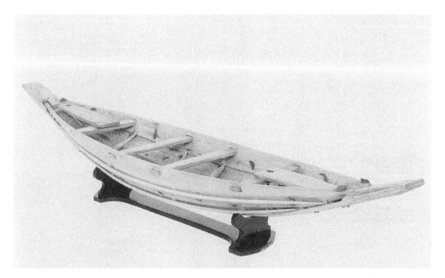

Figure 2.26 A nineteenth-century model of a *khelna* reverse-clinker boat from Khulna (Reg. No. 499/54). Photo: Science Museum, London.

Greenhill (1971: 104) in the 1950s and by Abrar Hossain in 1997. This model (labelled 'Khelna') went to the Victoria and Albert Museum in 1879 from the former India Museum, which had been set up by the East India Company in 1809. Although several iron nails have been used in the model, the Victoria and Albert Museum catalogue of 1880 states that the reverse-clinker planking of full-size boats was fastened with 'bamboo pegs'.

In addition to their reverse-clinker planking, the three types of Sylhet boat have a number of other features in common: they are built plank first; they have relatively long, narrow and shallow hulls; the stern is higher than the bow; their planking is laid (sometimes vestigially) in hulc fashion; and they have block or plank stems. The three types may be differentiated initially by reference to their hull shapes.

Type 1   *Sylheti nauka*: round hull in section; double-ended 'boat-shaped' in plan; medium L/B ratio.

Type 2   *Barki*: flat-bottomed with flaring sides in section; 'boat-shaped' in plan; medium L/B ratio.

Type 3   *Chhataki nauka*: near-rectangular in section; 'logboat-shaped' in plan; high L/B ratio.

There are also structural differences:

Sylheti nauka: Many reverse-clinker strakes, clearly laid hulc fashion; block-*goloi*; a moderate number of floors and crossbeams.

Barki: A few reverse-clinker strakes, truncated at the ends and not so obviously hulc-like; intermediate form of *goloi*; many lower crossbeams and some upper crossbeams;

Chhataki nauka: A few reverse-clinker strakes, truncated at the ends and not obviously hulc-like; plank-*goloi*; several foot-grip 'floors' and two crossbeams.

Subjectively, there also seem to be differences in the standards of boatbuilding, with the *Sylheti nauka* giving the impression of being craftsman-built with some personal touches and, at the other end of the spectrum, the *Chhataki nauka* that of being built of standardised parts.

It is tempting to see in these differences a progression from a hand-crafted, traditional, general-purpose boat (Type 1) to a simplified boat of lesser craftsmanship (Type 2) and on to the mass-produced boat (Type 3) designed for a specific function. This is not possible to demonstrate at present, but the hypothesis receives some support from the fact that the Type 1 and Type 2 boats are not unlike some of the boats that Basil Greenhill noted fifty years ago, although there is no clear one-to-one relationship. Greenhill's round-hulled, reverse-clinker Sylhet boats (his Class 2) ranged from expanded logboats extended by strakes (Greenhill, 1971: fig. 16) to large, fully planked cargo vessels (Greenhill, 1971: lower plate on p. 87). The boats observed by us in Sylhet seem to be somewhere in between these two ends of the spectrum. Visually, the *Sylheti nauka* and the *barki* bear some resemblance to two Sylhet reverse-clinker boats illustrated by Greenhill (1995A: respectively fig. 327 and fig. 105). However, there are differences. The first of these boats, although having hulc planking at the bow, had planking at the stern which was 'faired to the garboard strake' (see Fig. 2.12); while the second one is thought to have had a logboat base (Greenhill, 1995A: 254, 105).

If we look further back in time, the picture becomes even more opaque and resemblances to today's boats are more difficult to identify, although there is evidence for boats with different planking patterns at bow and stern, similar to that noted by Greenhill (Fig. 2.12): see, for example, the *pataily/pataeley* depicted by Solvyns in the late eighteenth century (Fig. 2.9) which has hulc planking only at the bow. Other vessels from this region of South Asia depicted by Solvyns (the *pettoua/petooa* and the *pansway/panswae*) and by Pâris in the early nineteenth century (the *pansway*; the *patile/pataily*; the *baulca*; and the *dinghi*) have hulc planking at both ends. Of these boat types, the *pettoua/petooa* (see Fig. 3.4 below) almost certainly had reverse-clinker planking, the *dinghi* possibly had it, while the *pataily/pataeley/patile* had overlapping planking which may have been European clinker. The worldwide evidence for reverse-clinker planking and hulc planking patterns is summarised in the Appendix.

One thing evident from this iconographic evidence is that both reverse-clinker planking and hulc planking patterns were known during the late eighteenth/early nineteenth centuries in the great river systems north of the Bay of Bengal. The *pansway* illustrated by Pâris may have had block stems (*goloi*), so this feature of the twentieth-century Sylhet boats may also have earlier roots. These main traits apart, there is little to link the Sylhet reverse-clinker boats of today with those depicted by Solvyns and Pâris. However, the resemblances between the Orissan/West Bengali *patia* of today and Solvyns' *pettoua/petooa* are stronger (see Chapter 3).

3

THE REVERSE-CLINKER BOATS OF
ORISSA AND WEST BENGAL

Seán McGrail, Lucy Blue and Eric Kentley

The generally emergent coastline of eastern India (Ahmad, 1972: 50–51, 180) is characterised by a series of prograding deltas dominated in southern Orissa by the 300 km arc of the Mahanadi delta (Fig. 3.1). This delta consists of three estuarine outlets: the Mahanadi, Brahmani and Baitarani rivers, which have joined together and protrude some 60 km to the east. Further north, and on into southern West Bengal, the coastal plain is narrower and deltaic progradation is less pronounced, although in the area of the rivers Boro Bulong and Subarnarekha rapid accumulation of silt has been noted since the sixteenth century (Deloche, 1994: 114; Patra, 1988: 151). In northern Orissa and southern West Bengal, alluvial deposits from the River Subarnarekha are transported to the northeast as littoral drift by the wind, forming beach ridges or barrier bars or spits. Further northeast, the coast of West Bengal is dominated by the silty fringes of the Ganges delta (Niyogi, 1968: 232, fig. 2). The predominant wind in this region is southwesterly from June to October and northeasterly during the winter months. Coastal currents largely correspond to these winds.

Work by Niyogi (1968), Sambasiva Rao *et al.* (1983) and Mahalik (1992) has afforded insight into the growth stages of the Orissan deltas, particularly in relation to the progression of ancient beach ridges (formed by wave action) and barrier bars (the product of littoral drift). Some of these ridges and bars, such as the Chandbali formations on the Brahmani-Baitarani delta, now extend out to 35 km from the coast and are believed to have formed over the last 6000 years. Conversely, a series of U-shaped dunes southeast of Konarak are estimated to have formed during the past 200 years or so, which would coincide with, and perhaps account for, the documented abandonment of Konarak (Sambasiva Rao *et al.*, 1983: 264–6; Pethick, 1984: 141; Patra, 1988: 204–205). Niyogi (1968: 236) established a comparable relative sequence of post-glacial growth for the Subarnarekha delta.

In recent centuries the process of coastal alluviation has become more rapid due to a combination of factors, including the tectonic uplift of the

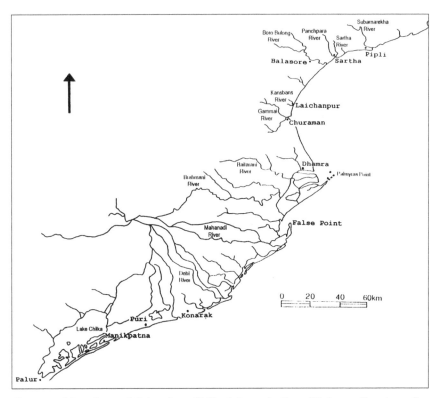

Figure 3.1 Map of coastal Orissa from Chilka Lake to the Bay of Balasore. Drawing: after Deloche 1994: 115, fig. xxi.

coastal belt, which has caused the emergence of coastal dunes. The formation of these dunes has also been encouraged by the northerly-flowing longshore current, a product of the SW monsoon, which deposits sand bars across river mouths, cutting off many of the riverine ports.

Monsoonal floods add to the problems of alluviation, particularly during the cyclone season when the river basin catchment funnels large volumes of water into the delta, causing the rivers to silt and choke behind the coastal sand ridges. The increased sediment load is also a product of the expanding practice of deforestation and subsequent erosion in the uplands (Patra, 1988: 203) and has had a disastrous effect on the river systems of the Subarnarekha delta, resulting in increased sediment loads. All these factors have contributed to the accelerated levels of alluvial deposition and the prograding nature of the coastline.

The deltaic flat (or plain) associated with the Subarnarekha estuary is now generally *c.* 3 km wide, and to seaward of it there are a number of barrier bars or spits (Niyogi, 1968: 232). Towards the northeast this deltaic flat narrows and, around 3 km west of Digha, it disappears completely and a ridge of dunes, some 200 to 300 m wide, dominates the coastline. The dunes run parallel to the shore and vary in height from 4 to 15 m. Inland of them is a marine terrace of silty clay, while to seaward is a sandy beach c. 100 m wide which is open to ocean waves. The foreshore and sea bed slope gradually: depths of only 4 to 6 m (2 to 3 fathoms) at 2 km (c. 1 nautical mile) from the shore. The open nature of the shoreline, the onshore wind and the low declivity of the beach result in the generation of surf with spilling breakers (Van Doorn, 1974: 228–39) and erosion of the coastal dunes.

The stretch of coastline between the River Subarnarekha and the River Champa is a lee shore with breakers and is thus different from the coast to the northeast and to the southwest. Propulsion by oar rather than sail is best in such surf and wind conditions, and using a seine net from the beach becomes the most practical method of fishing.

The maritime history of this coast

The economic significance of the northwestern coast of the Bay of Bengal and the importance of its ports as a commercial link between East and West is attested from at least Roman times (Casson, 1989; Mohapatra, 1983; Parida, 1994; Ray, 1994: 48–52), and there is a wealth of historical evidence for the extensive use of this coastal hinterland from the sixteenth to the nineteenth century AD (Patra, 1988: 19–77).

From the second century AD this region, like many other coastal regions of South Asia, supplied spices, diamonds and precious stones to the Roman Empire (Patra, 1988: 10). Ptolemy (*Geog.* VII.1.16) named several Orissan river ports of this period, including Jaganatha-Puri, Katak, Konarak, Pipli/Balasore and the mouths of the rivers Mahanadi, Brahmani, Baitarani (?) and Subarnarekha (?) (Deloche, 1994: 112–113).

During medieval times, c. twelfth to mid-sixteenth century, earlier trade links with Southeast Asia, Ceylon and beyond to China were expanded (Behera, 1994: 58), with the export of commercial crops, industrial products such as textiles and perfumes, diamonds and ivory (Patra, 1988: 10–11; Pattanayak and Patnaik, 1994). The expansion of trade during this period encouraged the increased use of inland waters to transport grain and other products from the hinterland and to move salt inland from the coastal margins. Boats are depicted in temple sculptures and on palm-leaf inscriptions of this period (Pattanayak and Patnaik, 1994: 8; Behera, 1994: 67).

Early ports in river mouths and within coastal lagoons (Fig. 3.1) are known, from local inscriptions and foreign accounts, to have been established at Chelitalo, Khalkatta, Dhamra, Puri and Konarak (Deloche, 1983: 444; Patra, 1988: 10–13, 204–205; Pattanayak and Patnaik, 1994: 8; Parida, 1994). However, only two coastal settlements, Manikpatna and Palur, both of the early historic period and situated in the Chilka region, have so far been excavated.

Towards the end of the medieval period the economic prosperity of the region fell into decline, partly due to competition from Arabic, Portuguese and Dutch traders and partly due to the threat of Muslim invaders. This, coupled with increased occurrence of natural disasters such as drought, famine and cyclones, the gradual siltation of the waterways and the progradation of the coastline eastwards, brought about a reduction in maritime commercial activities and rendered many of the ports obsolete, isolated some kilometres inland or no longer accessible from the sea (Pattanayak and Patnaik, 1994: 9–10).

It was not until the sixteenth to nineteenth centuries, during the time of European expansion, that maritime activities were revived and important centres of maritime commerce and shipbuilding were established along the coast. In 1514 the Portuguese established a base at Pipli, some 6 km from the mouth of the River Subarnarekha (Patra, 1988: 152). Around 1625 the Dutch established an important commercial port at Balasore which was eventually to replace Pipli, after flooding of the Subarnarekha washed away that port and a hazardous sandbar began to form at the river mouth (Patra, 1988: 103; *Bay of Bengal Pilot*, 1887: 161). The harbour established at False Point (Fig. 3.1) was said to be the best harbour along the coast and provided access to the Mahanadi delta; while the port at Dhamra guarded the estuaries of the Baitarani and Brahmani rivers. The port of Sartha, situated at the junction of the Panchapara and Sartha rivers, to the south of the Subarnarekha, was served by an open roadstead. Churaman was established on the River Gammai and, to the north, the smaller port of Laichanpur was developed on the River Kansbans (*Bay of Bengal Pilot*, 1887: 157; Deloche, 1994: 116).

Many different boats were noted on the Bay of Bengal coast from the mid-sixteenth century onwards. They included the flat-bottomed *patilla*; the *purgoa*, used on the Hueghli and at Balasore and Pipli to load and unload large ships; the *malangi*, used for transporting grain to inland ports up river; and the small *holah*, which carried salt and textiles to Cuttack via inland waters (Patra, 1988: 19, 153; Toynbee, 1873: 93). With the increase in commercial activities, shipbuilding expanded at Pipli and Balasore, using timber from the forests of Sambalpur, Balasore, Cuttack and Ganjam (Patra, 1988: 69–76). However, throughout this period of maritime expansion, no attempt was made to prevent the rivers from silting, to dredge the sandbars that increasingly formed at river mouths or to buoy the channels (Patra, 1988: 183), and thus the decline of the ports was inevitable.

These recent environmental changes have led to the abandonment of many important Orissan harbours and there has been a subsequent reduction in the number of river craft. However, traditional boats continue to be used by fishermen in estuary mouths and in coastal waters.

70

To date, archaeological discoveries are restricted to a chance find of timber from a sixteenth-century barge (most probably of European design) in an abandoned tributary of the River Boro Bulong at Olandazsahi, near Balasore (Behera, 1994: 67). Ironically, the alluvial processes that brought about the decline in maritime activity also provide the means for its survival in the archaeological record. Despite the otherwise adverse tropical climate, which would usually result in the rapid deterioration of organic material, the anaerobic nature of the deltaic alluvium can preserve boat and ship timbers and the lower elements of waterfronts.

Early evidence for reverse-clinker vessels

A twelfth-century stone relief in the Jagannath temple at Puri, Orissa, is only accessible to Hindus, but it has been reported that a ceremonial barge carved in relief in black granite on the wall of the Jagamohana building has reverse-clinker planking depicted (Deloche, 1996: 206, fig. 4d). Reverse-clinker planking is also depicted on two other eleventh/twelfth-century stone carvings thought to be from Orissa: one held in the Victoria and Albert Museum in London (Fig. 3.2; see Guy, 1995) and one in the India Museum in Calcutta (Fig. 3.3).

Six hundred years later, Frans Balthazar Solvyns (1799) illustrated a *pettoo-a* with reverse clinker (Fig. 3.4) in volume 3 of his book *Les Hindoos* (see

Figure 3.2 Stone sculpture of a boat (235 mm in length) with reverse-clinker planking, thought to be from eleventh/twelfth-century Orissa. Victoria and Albert Museum No. IS.475-1992. Photo: Victoria and Albert Museum, London.

Figure 3.3 Stone relief from Orissa dated to the eleventh/twelfth century AD. Photo: Indian Museum, Calcutta.

Hardgrave, 2001). Although this *pettoo-a* does not have the high rising ends of the twentieth-century *patia* (Fig. 3.5), and the eighteenth-century vessel was a cargo ship rather than a fishing boat, there are clear resemblances between the two vessels. There is, however, no clear link between them and the eleventh/twelfth-century depictions, although they all have reverse-clinker planking in common. The problem of the interpretation of iconographic evidence for reverse clinker and the associated problem of 'hulc planking', and the impact of both these features on European medieval maritime studies, are considered in more detail in the Appendix to this volume.

Fieldwork

During February 1996, fieldwork was undertaken in coastal northern Orissa (Fig. 3.6) with the aim of locating and documenting *patia* reverse-clinker boats which had been noted some years earlier by Mohapatra (1983) and by Professor B. Arunachalam (Deloche, 1994: 170, fig. 35). There are three types of *patia* today: a boat propelled by sail or oar and used for estuary and coastal fishing; an oared boat of about the same size used for beach seine-net

72

Figure 3.4 Late-eighteenth-century drawing of a *pettoo-a* 'from Balassora or the coast of Palmira', by Solvyns. Photo: Bodleian Library, University of Oxford: F.C. 4(10) Section 9, No 5.

fishing; and a larger motorised version used for off-shore fishing. Research was concentrated on the traditional *patia*: the sailing and the oared boats.

Patia reverse-clinker sailing boats were found at two sites: at Kasafal, a beach market on the southwestern bank of the River Panchpara, northeast of Balasore; and at Talesari, a beach landing place further east, on the northern shore of the Subarnarekha estuary and close to the border with West Bengal (see Fig. 3.6). At both these sites several boats were studied and their crew interviewed, and at Kasafal a small sailing *patia* was selected as representative of *patia* in that region and was recorded in detail (Fig. 3.5): she measured 6.88 x 1.32 x 0.75 m. A short coastal passage was subsequently undertaken in this boat.

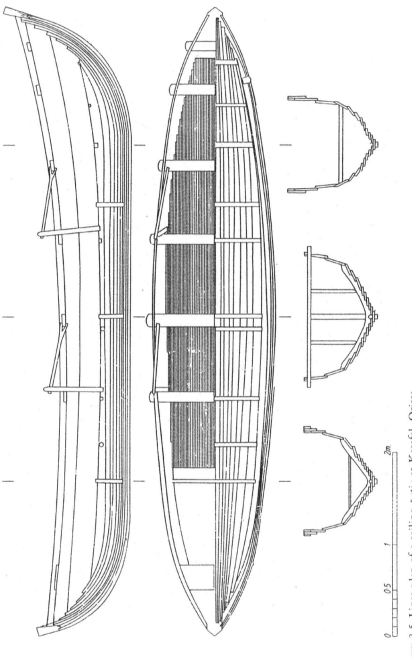

Figure 3.5 Lines plan of a sailing *patia* at Kasafal, Orissa.

0   0.5   1   2m.

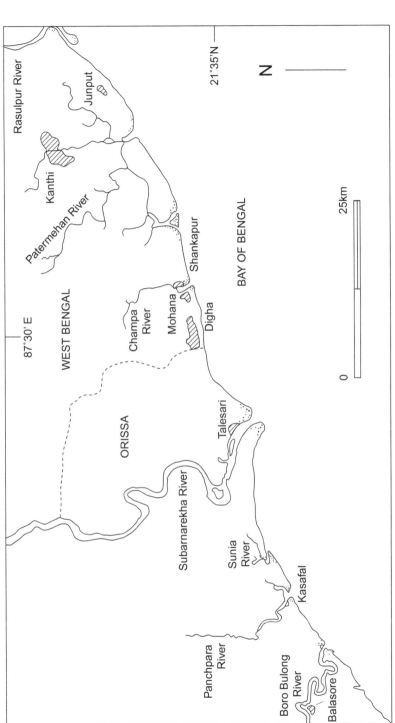

Figure 3.6 Map of coastal northern Orissa and southern West Bengal showing the main rivers and the sites where *patia* fishing boats were found in 1996 and 1997.

At Talesari an itinerant boatbuilder was building one of the larger *patia* and repairing another on an informal site set back from the foreshore and protected by dunes. Here it proved possible to record features which had been masked by superimposed fittings or hidden under a layer of tar on the working boats and to question the builder about his boatbuilding techniques and his tools. When this site was revisited in 1997 (see below), it was found that the protecting dunes had been breached and eroded. A surface search, without removing vegetation, of the area of the building site revealed no signs of the earlier boatbuilding activity.

In November 1997 we returned to the border area of northern Orissa/ southwestern West Bengal with the aim of mapping the northerly distribution of this boat type and of documenting any significant variations in boatbuilding and boat use. Oared *patia* were found at Digha, all of them berthed above high water since this was not the usual season for seine-net fishing. Four *patia* in an area protected by sand dunes were examined in some detail and a representative transverse section was measured.

About 1 km inland of Digha and at some height above sea level, in the village of Gangatharpur, an oared *patia* was being built on a cleared site near the builder's house. The builder, Nirenjan Das, was not working at that time, but he allowed us to examine his boat and described his building methods to us.

Distribution of the patia

In the early 1980s a great many *patia* reverse-clinker boats were to be found in northern Orissa. Kalavathy and Tietze (1984, table 5) published the numbers of *patia* noted in each district in this period: see Table 3.1.

During the 1996 period of fieldwork we learned that *patia* were no longer used in some of these districts and that the southern limit of the *patia* was then at or near Kasafal. During the 1997 fieldwork several West Bengal fishing villages and landing places were visited from the Orissan border as far

Table 3.1 Survey of *patia* in the 1980s

Orissan district	Number of boats
Dhamara	11
Basudevpur	272
Gopalpur	217
Balasore	7
Baliapal	99
Bhograi	181
	787

north as Junput (see Fig. 3.6). Digha, just into West Bengal, proved to be the most northerly/easterly point where *patia* can now regularly be found. Only a single *patia* (a motorised one) was seen beyond Digha: at Mohana, a fish landing place and market about 2 km to the east. West Bengal has a flourishing fishing industry and a wide range of boat types, however the *patia* is firmly regarded as of Orissan origin. Indeed the family of the only *patia* builder the authors could locate in West Bengal had migrated from Orissa two generations previously, and *patia* seem to have been built in Digha only since *c.* 1947.

Although there are significant numbers of *patia* at Kasafal and at Digha, the largest concentration appears to be at Talesari: on a single afternoon in November 1997, 118 were counted. At Kasafal there are sailing and motorised *patia*, and the latter predominate. At Talesari the dominant type is the sailing *patia* (of the 118 seen, 72 were small sailing boats, 40 were motorised and 6 were oared). At Digha, where there is a surf-ridden foreshore, there are only oared *patia*.

The motorised and the sailing *patia* use drift nets: the sailing boats fish out to about 5 km, the others out to 20 km. The oared *patia* use seine nets from the beach. At all three sites – Kasafal, Talesari and Digha – other types of craft were also in use, including motorised vessels of some size as well as traditional boats propelled by oar or (but not at Digha) by sail.

Size and shape of patia

McKee (1983: 79) classified a boat as 'deep' where the beam/depth ratio is less than or equal to 2.00 and 'narrow' where the length/beam ratio is greater than or equal to 3.75. Sailing *patia* have shape ratios of B/D = *c.* 1.76 and L/B = *c.* 5.11, while oared *patia* have B/D = *c.* 1.67 to 1.56, L/B = *c.* 4.5 to 4.68. The traditional *patia* can therefore be described as both narrow and deep.

The oared *patia* measured at Digha were between 7.74 and 8.52 m in length, with beam measurements from 1.72 to 1.82 m and depths of 1.03 to 1.17 m, and they had 14 or 15 strakes each side. In contrast, Orissan sailing *patia* were around 7 m in length with 11 strakes, while motorised ones were 8.5 to 10.5 m long and had 14 to 16 strakes (Mohapatra, 1983: 5). The Digha oared *patia* were thus intermediate in size.

Patia lower planking sweeps up at bow and stern to end above the waterline, not at a post but on an angled line (Figs 3.5, 3.7), similar in this respect to representations (Figs 2.5, 3.8) of the medieval European hulc (see Appendix). The space between the two high-rising groups of reverse-clinker planking is filled by three or more broad strakes so that the boat generally has a moderate sheer (upward curve of top edge of sides), more at the bow than the stern. Most of the oared *patia* seen at Digha had a comparable sheerline, but two older, disused boats had only a slight sheer at the bows.

Figure 3.7 The bows of an oared *patia* at Digha, West Bengal. The reverse-clinker
planking is laid 'hulc fashion', rising at the ends to an angled line. European
clinker strakes 'fill the gap' between the two ends.

Building the *patia*

Patia are built almost entirely of *sal* (*Shorea robusta*) and are heavily coated
inboard and outboard with tar. In West Bengal, when *sal* is unavailable,
arjun is used; *babla* is used for the framing. Apart from modifications made
to accommodate the engine (principally additional strengthening timbers
and a skeg to take the propeller shaft), all sizes of boat are evidently built
in the same manner. Some of the features of the *patia* have no analogue in
European boatbuilding and there is no term in the English language for
them. The nearest available term has therefore been chosen: these are
illustrated in Figure 3.9.

Figure 3.8 Seal of New Shoreham, Sussex, of *c*. 1295. The vessel depicted has hulc planking which may be laid in reverse-clinker technique. Photo: National Maritime Museum, Greenwich.

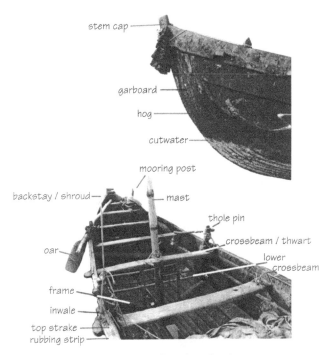

Figure 3.9 Diagram illustrating terms used to describe the *patia* structure.

Design

No drawings, models or moulds (templates) are used when building a *patia* and, although a centreline is set up and the builder has a measuring tape calibrated in inches, measurements are used only to a limited extent, mainly to ensure symmetry. The boats are largely built 'by eye' and much depends on the experience of the builder, but there appears to be reliance on rules of thumb for certain aspects. The key to the 'design' of the *patia* appears to lie in memorised data which determine the effective breadth of strakes and the angles of bevel on the strake edges.

The Digha builder, Nirenjan Das, quoted to us a rule of thumb giving the laps for the 10 or 11 reverse-clinker strakes of a 'standard' size of *patia*, which was said to be 39 ft (11.89 m) in overall length (longer than any *patia* actually observed at Digha). The strakes were said to be fashioned from planks which were 6 inches broad and 1 inch thick (152 x 25 mm). Each strake was laid on the inboard face of the strake below so that its lower edge was a fixed distance from the lower edge of that lower strake. These distances, which, at any one station, were said to be the same for all strakes, are given in Table 3.2.

Since the strakes are all a standard breadth and, at any one station, the overlaps are standard, the effective breadth of every strake at any one station is the same. Thus, since Nirenjan Das's table does not include information on bevel angles, the midships section of the idealised boat described by Das is as in Figure 3.10 (lower). The lower part of this section, planked in reverse clinker, gives the boat a V-shaped hull, whereas an actual boat recorded at Digha (Fig. 3.10 upper) had a more rounded hull, due not only to bevels being worked along at least some of the plank edges but also to the use of varying breadths of planking. It seems clear, therefore, that Das's numbers are, at best, a guide (probably of most use towards bow and stern). The *patia* is built 'by eye': judgement, based on experience but possibly on other (unrevealed) 'formulae', is used by the builder to select the breadth of plank, the bevel angle and the overlap needed to achieve the shape of hull required.

When asked about the reasons for choosing to use the reverse-clinker technique, and its perceived advantages over other methods, the Digha

Table 3.2 Formula for determining strake laps of an oared *patia*

Station	Distance from lower edge of strake below
Amidships	3½ in. (89 mm)
At 12 ft (3.7 m) from one end	3 in. (76 mm)
At 3 ft (0.9 m) from one end	2½ in. (64 mm)
At 2 ft (0.6 m) from one end	2 in. (51 mm)
At 1 ft (0.3 m) from one end	1 in. (25 mm)

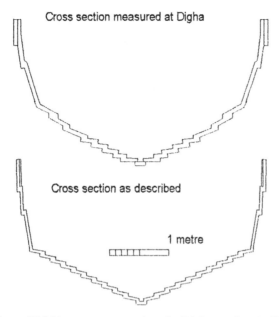

Figure 3.10 Above: Midships transverse section of a Digha oared *patia*. Below: Idealised *patia* transverse section described by the Digha boatbuilder.

boatbuilder gave similar answers to those given by the Talesari builder: 'this is the natural way to lay the planks'; 'we have always done it this way'; reverse clinker is 'good for seakeeping and for fishing', it allows the boat to 'cut through the water'.

The building sequence

This description of the building of a *patia* is mainly based on the boat selected for recording at Kasafal, on observations made on other similar sailing *patia*, and on an interview with the itinerant boatbuilder at Talesari.

The 'backbone'

Building begins with laying the plank-keel which, on the sailing *patia*, is *c*. 55 mm sided (broad) by 30 mm moulded (deep). On top of this a thin plank (approximately 115 mm by 22 mm), or series of planks scarfed together, is nailed, forming a hog (see sections in Fig. 3.5). More planks are scarfed to the ends of this plank, so that the hog runs beyond the keel and sweeps up to form stem and stern posts; this 'extended hog' constitutes the backbone of the whole boat. What may be called stem and stern 'caps' are scarfed to the top of each post, capping post and adjacent 'cutwater' (see below).

Half-rounded timbers are fastened onto the outboard side of the hog, between the keel and the cap, at both ends. These are of little structural importance: they are not posts but perform the function more of a cutwater, streamlining the water-flow along the hull. Both cutwaters are fastened to the keel in vertical half-lap scarfs.

The planking

The edge of the first strake (the garboard) is laid on the upper side of the hog; then the lower side of the second strake is laid inboard of the upper side of the garboard, and so on – a perfect example of plank-first (shell) building. Planks within strakes were generally scarfed together in a vertical lap, although some above the waterline appeared to be plain butted. The process of planking-up continues until the eighth strake (the last in reverse clinker) is fitted and fastened. The lower edge of a plank is precisely shaped before it is fastened to the strake below; its upper edge is shaped and bevelled *in situ*.

In shape, all eight strakes gently taper for most of their length, but about 500 mm from both ends of each one they abruptly broaden on the outboard side (Fig. 3.11). Each spatula-shaped strake end forms a landing to which the end of the next higher strake of reverse-clinker planking can be securely

Figure 3.11 A view inside the unfinished bows of a large *patia* at Talesari. The ends of each strake of reverse-clinker planking are broadened to provide a landing for the next higher strake.

fastened. The eight reverse-clinker lower strakes form the bottom of the *patia* hull and sweep up to form gracefully rising ends (Fig. 3.12).

The ninth strake stands on the edge of the eighth strake, forming a relatively hard chine – where the bottom of the boat meets the side of the boat. This is approximately the loaded waterline amidships. The lower edge of the tenth strake is fastened *outboard* of the ninth strake, and the eleventh strake is similarly fastened outboard of the tenth: that is, the tenth and eleventh strakes are fastened in European clinker fashion and not in reverse-clinker fashion.

A broad inner rail, of the same thickness as the strakes, is fitted on the inside of the eleventh strake, flush with the sheer. It extends down to the top of the tenth strake, giving the appearance of flush-laid planking. A simple rubbing strake is fitted to both sides of the boat aft of the sixth thwart.

All the timber for strakes at the Talesari boatbuilding site was 22 mm thick (i.e. one inch nominal). On the Kasafal boat, there appeared to be some variation in strake thickness, but the tar coating made it difficult to take precise measurements. Even with a thickness of 22 mm, the wood will not bend naturally into the shapes required. Each plank is heated over a fire and bent to shape with a simple weight and lever device.

The ninth, standing, strake is fastened to the eighth (uppermost reverse-clinker) strake by spikes which are driven through the eighth into the lower edge of the ninth. Strakes laid European clinker fashion are fastened as in Europe (see Fig. 2.3, upper). Reverse-clinker strakes, on the other hand, are

Figure 3.12 The stern of a large *patia* being built at Talesari. Note the graceful, up-sweeping curve of the reverse-clinker planking in the lower hull.

fastened together by driving an iron nail through the lap from *inboard*. The point of the nail protruding outboard is then turned through about 90 degrees with pliers and hammered back into the planking to clench the nail by hooking the point through 180 degrees, giving the appearance of a horizontal staple. Nail spacing is approximately 150–200 mm amidships, closer towards the ends. Once construction is finished, raw cotton is forced into the seams as a caulking and the seams are tarred.

The framing

As can be seen from Figures 3.5 and 3.9, framing in the sailing *patia* consists of floor timbers, crossbeams and side timbers. There are just three floors, from eighth strake port to eighth strake starboard, joggled to fit the reverse-clinker planking, with limber holes to allow the free flow of bilge water. The aftermost floor also incorporates a dwarf bulkhead. These floors are fastened to the planking by nails driven from outboard and hook-clenched inboard.

There are four side timbers, one forward of the fourth and sixth crossbeams (counting from the bow) to port and the third and fifth beams to starboard. They stand upon the eighth strakes and are joggled to fit tightly against the ninth to eleventh strakes, to which they are nailed. Each of these side timbers extends some 200 mm above the sheerline so that it can be used as an oar pivot or a mooring bitt.

The seven crossbeams, broad enough to be used as thwarts, sit on top of the tenth strake and are clamped into place by the eleventh strake, which is notched to take them. Oarsmen can sit on the third, fourth and fifth crossbeams using the fourth, fifth and sixth crossbeams as stretchers. The mast is made fast to the after side of the fifth beam. The aft-most crossbeam forms a platform on which the steersman can stand. All crossbeams, except this one, extend outboard, providing handy purchases when manhandling the boat. Protruding crossbeams are not a feature of just the *patia* – they are characteristic of Indian boat design in general.

The framing of the oared *patia* differs from that of the sailing *patia* in three main ways:

(1) The oared *patia*'s crossbeams, which are used as thwarts, are grouped forward of amidships, leaving a clear space aft for the seine net. The central beams are sometimes arched and are supported underneath by pillars.
(2) The oared *patia* has six floor timbers, rather than three.
(3) While some oared boats have the simple framing of the sailed *patia*, others have composite frames with side timbers overlapping floors. These side timbers extend above the top of the sides and are used as oar pivots with angled struts as in the sailing *patia*. Overlapping timbers are not fastened together. The separate side timbers of those oared *patia* with

the simple framing of the sailing boats do not extend above the sheer, and additional side timbers, without struts, are added to be used as oar pivots. One oared boat was noted, however, with the unusual combination of simple framing and oar pivots with struts.

From these observations it would seem that the Digha boatbuilder has been experimenting to find the combination of framing best suited to beach seining. Apart from these differences in framing, which appear to be associated with different uses of the boats, and the fact that the sailing *patia* has a somewhat sharper underwater hull in section, the Digha oared and the Orissan sailed *patia* are similar in shape and structure.

'Decking'

As the *patia* is narrow and deep, walking in the bottom can be both precarious and uncomfortable. A series of split-bamboo mats are therefore laid at the level of the top of the eighth strake and supported by four fixed lower crossbeams and by three movable bamboos, jammed between the sides of the boat.

Decoration and ritual

The *patia* is said to be regarded by fishermen as a living spirit that not only provides them with a livelihood but also protects them from danger. In the Cuttack and Puri districts (southern Orissa) boats are identified with the goddess Mangala, whereas in Kasafal, Talesari and Digha, where Bengali influence is strong, boats are associated with Kali. To represent her, the forward-facing faces of the *patia*'s stem cap are inscribed with eyes, nose and mouth representing a stylised head (Fig. 3.13). A tiered apron of brightly coloured cloths, or a painted representation of one, is tied around the base of the cap. Each tier is of a different pattern, and tinsel and/or flowers are often added. At Digha there was additional painting on the bow planking just below the sheer: usually this was in rows extending the tiered effect of the apron, but sometimes a hooked tool or weapon, a symbol of Kali's might, was depicted.

A full figure of Kali is carved on the starboard planking, inboard, roughly amidships. This figure is covered with vermilion, and incense sticks are pushed into it before the boat sets sail. This carving is more crudely formed than those on the stem cap, which are very finely executed.

The inboard face of the stem cap and all the faces of the stern cap are decorated with geometrical shapes which vary from boat to boat. Generally these decorations consist of a half-circle inside which are arcs, stars and other symbols. Although informants were explicit about the role of Kali in the

Figure 3.13 A *patia* stem cap inscribed with the eyes, nose and mouth of Kali and
decorated with a tiered apron.

boat's decorative scheme, they appeared to attach no special significance to
these other patterns.

At Digha several boats had white patches of rice-dust and water which had
been ceremonially applied to the outer face of the side planking in three
places. The only other decorative feature observed on Orissan or Bengali *patia*
was the two-dimensional finial shaping given to the starboard mooring bitt.

A model of an oared *patia*

Model building is not part of the *patia* design and building process, and no
toy boats were observed. Nevertheless, the Digha boatbuilder Nirenjan Das

readily accepted a commission for a model *patia*. Apart from the requirement that the model should be easily carried by one person, no formal specification was given to him.

The result, completed and delivered to Calcutta within four days, is illustrated in Figure 3.14 and is clearly based on an oared *patia*. As happens in cultures where model building is not part of the design process, the proportions are inaccurate, but the characteristic features are present.

The model is 500 mm in length overall, 442 mm between the inner faces of the 'posts'; the beam is 136 mm to the inside of the planking and 160 mm overall; the depth is 110 mm. These dimensions give unrepresentative shape ratios: the model is much broader and deeper than the real boat. Furthermore, the model has a greatly exaggerated keel, carved as a single piece with the cutwater. Nevertheless, much of the detail is correct: for example, the stem and stern post caps are shown as separate pieces, and the forward cap even has a simplified representation of Kali carved on it.

There are seven reverse-clinker strakes, four European clinker strakes and a rubbing strake each side. This planking, *c.* 4 mm thick, is not fastened in the manner of the real vessel but with what appear to be panel pins driven from inboard. There are seven crossbeams, including a broad one at the stern for the steersman. The third and fourth beams from the bow are each supported by three pillars fitted into associated floor timbers. These two floors, which appear to be joggled, are the only ones on the model, although there are slight lower crossbeams in the bow and stern.

The model has four oar pivots: on the port side, aft of the second and fourth crossbeams; on the starboard side, aft of the third and fifth crossbeams. A single, one-piece oar was supplied, 216 mm in length, with a blade measuring 90 x 25 mm. Although this oar is out of proportion, it has

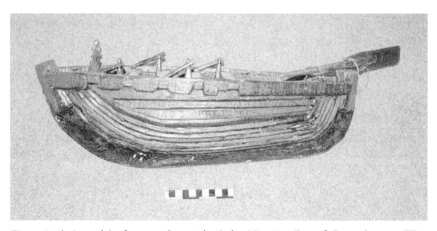

Figure 3.14 A model of an oared *patia* built by Nirenjan Das of Gangatharpur, West Bengal. Three of the four oar pivots can be seen, and there is a mooring bitt in the bows and a steering oar protruding over the stern.

been fashioned with the correct cross-sectional shapes. The model steering oar, which is 320 mm long, with a blade of 170 x 32 mm, lies naturally to port of the 'post' in a string grommet through a hole in the stern cap.

The model is finished with a generous coating of tar. The oar handle is coloured red (although no real oars were seen like this), and the two caps and a point on the starboard planking (corresponding to the position of an image of Kali in a boat) have been covered with red pigment. This may be evidence of a ceremony to mark the completion of the model: when delivered, there were flowers inside the hull.

Boat operations

The sailing patia

The crew

The crew of the boat measured consisted of two men and a boy. The older of the two men was 'in command' of the boat but pulled an oar when needed; the younger one was helmsman and also sailing master; the boy pulled an oar, towed, bailed out (using a plastic container) and moored the vessel.

Under oars

Two oarsmen sit amidships on the third and fourth crossbeams, with their bare feet braced against the fourth and fifth crossbeams (Fig. 3.15). They each pull an oar pivoted in a rope grommet which is attached to a side timber forward of the fourth and fifth crossbeams. Each side timber is supported at its upper end, near the grommet, by an angled spar, the forked lower end of which takes against the crossbeam next forward, i.e. third or fourth. These spars are lashed to the top strake through a hole in the planking. The loom of the oar, within its grommet, takes against a groove in this forked timber. A third oar could also be used by an oarsman sitting on the fifth crossbeam and pivoting his oar in a grommet forward of the sixth crossbeam.

The oars, *c.* 3.30 m in length, each consist of a natural spar, 2.95 m long, to which a partly overlapping blade, 0.85 m in length, is bound with nylon cord. The blade has a near-aerofoil section and a breadth at its tip of 135 mm.

Under sail (Figs 3.16 and 3.17)

The mast, a 5.08 m bamboo spar with a lower end diameter of 70 mm, has a metal ring near the upper end to take the halyard. There is no mast step, but the mast stands directly on the hog, near amidships. It fits into a semicircular recess cut out of the leading edge of the mast beam (fifth from forward) and

Figure 3.15 A sailing *patia* under oars off Kasafal: two oarsmen, each pulling a single oar.

is held in place by a grommet, which is tightened by a Spanish windlass or 'tourniquet', the wooden bar of which is lashed by codline to the mast, or to the mast beam, or to the central of the three pillars which support this beam. There is one shroud, which is fastened to the protruding end of the mast beam, port or starboard, whichever is to windward.

The yard is a 3.38 m bamboo spar of 38–40 mm diameter. The broader end has a short projection, some 32 mm in diameter, and is always before the mast. The halyard is bent to the yard at a point which is between one fifth and one sixth of its overall length from its forward end. This means that, with the sail bent to the yard, approximately one eighth of the sail area is forward of the mast when on a reach or tacking. The crew expressed this relationship by reference to the head of the sail: one cubit is before the mast with seven cubits abaft.

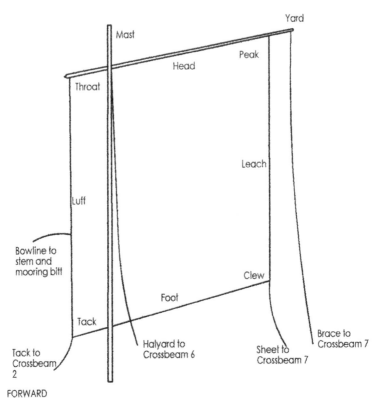

Figure 3.16 A diagram, not to scale, of the *patia*'s sailing rig, showing the terms used in the text. For clarity, the single shroud which runs from the mast head to the windward end of the mast beam has been omitted.

The sail is made of 17 rectangular, woven polypropylene sacks of varying sizes, with reinforcements at the corners, all sewn together in double hems. The head and the foot are both 3 m in length, the luff (leading edge) is 3.70 m, the leach (trailing edge) 3.50 m, giving an aspect ratio of *c.* 1.2:1. The luff is thus slightly longer than the leach, but as the sail always seems to be set peaked (i.e. with the after end of the yard higher than the forward end), this difference is not readily apparent. A bolt rope is neatly stitched all round the edges, and the sail is laced to the yard through gaps in this stitching along the head. The sail is positioned on the yard so that its throat (forward upper corner) is at the leading end of the yard while the peak (after upper corner) is some 38 cm from the after end, thereby leaving this after portion of the yard without sail.

Once the sail is hoisted, the halyard is used as an 'angled backstay' by securing it to the windward projecting end of the sixth beam. The tack (from the forward lower corner of the sail) is taken either to the second crossbeam

Figure 3.17 A *patia* with sail set while beached at Kasafal. Compare with the diagram of the sailing rig in Fig. 3.16.

or to the nearby mooring bitt. The sheet (from the after lower corner) and the single brace (from the after yardarm) are taken to the seventh crossbeam, near the helmsman. A bowline runs from the luff, about $\frac{1}{3}$ up from the tack, to the stem and thence to the mooring bitt. There are no reef points or other means of shortening sail.

From its shape, the sail is probably best described as a form of lug. However, the yard is not dipped aft of the mast, like a lug, to shift the sail from one side to the other but is handled more like a lateen. The halyard/angled-backstay is first eased out; the after end of the yard, with the peak of the sail, is next brought down and forward of the mast, then aft and upwards,

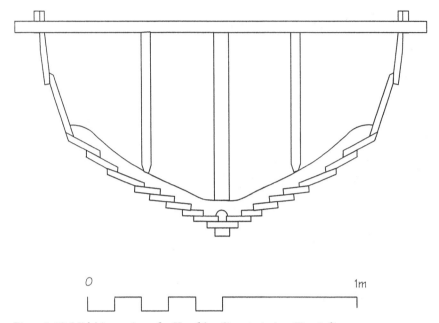

Figure 3.18 Midship section of a Kasafal sailing *patia* (see Fig. 3.6).

on the new leeward side of the boat. The single shroud, the single brace and the sheet are shifted to the other (windward) side of the boat. During a two-hour voyage – down the estuary of the River Panchpara, out to sea and return – the wind was only 'light airs', followed by a slight sea breeze, so the tacking ability and leeway of, and any helm carried by, the *patia* could not be investigated. However, the relative sharpness of the lower hull in cross-section (Fig. 3.18) and the fact that a bowline is fitted to the luff of the sail (see Fig. 3.16) suggest that this boat has some windward ability.

Towing

During the return to the estuary mouth on the early flood tide, which was not the fishermen's normal practice, banks to the south of, and parallel to, the channel were still dry. As only slow progress was being made under sail, the youngest crew member was sent ashore with a line from forward to tow the boat until the water became too deep for him.

Steering

The steering oar was 3.51 m overall and consisted of a natural spar, 3.20 m long, to which a blade of aerofoil section was scarfed with an overlap of 0.46 m. A check line led from the upper part of this blade to the top of the stern post. The steering oar was pivoted in a large grommet attached to this post and led through a second, smaller grommet, the one to port of the post or the one to starboard, depending on the position of the sail: for example, on the port tack (sail to starboard) the steering oar was to starboard. The steersman either sat, well aft, on one gunwale with his feet on the other and gripped the oar in mid-loom or, when more leverage was required, he stood on the seventh crossbeam and used the full length of the loom. The boat was turned by moving the loom to port or starboard; heading was maintained or 'fine tuned' by rotating the loom about its long axis by twisting the wrists.

Mooring

The boat can be anchored by a five-prong grapnel with a line to the second crossbeam and to the mooring bitt. The yard, after being slipped from the sail lacing, was sometimes used as a portable mooring post in the shallows. After taking the ground, the boat was propped upright by jamming portable crossbeams under the protruding parts of the fixed crossbeams. Alternatively, an oar was used with its blade lashed to the mast.

Fishing

The sailing and motorised *patia* are used to set pelagic drift nets, generally in the mornings, the main season being from September/October to March/April. The sailing *patia* work within about 5 km of the shore, while the motorised may go out to 20 km. All nets observed were of man-made materials (probably polyethylene). Mohapatra (1983) distinguished, by overall size and by mesh size, several different types of drift nets used by *patia*. However, all boats seem to catch principally clupeids (members of the *Clupeidae* family of oily flat fish, including sardines and herrings), particularly shad.

Nets are carried in the bottom of sailing *patia*, in the sternsheets between the dwarf bulkhead and the aftermost crossbeam. Wear marks on the top edges of the rubbing strake and top strake, created by hauling in full nets, were clearly visible on several boats. A Dan buoy was carried for use as a floating marker: this was a natural spar with a plastic can at the waterline, a brick at its lower end and a piece of red material at the top.

The oared patia

The oared *patia* at Digha and Talesari are used exclusively for beach seining. Seine-net fishing from a beach is one of the most labour-intensive fishing activities along the Indian coast and, generally speaking, it is undertaken by the poorest people in a locality. At Digha, seine-net fishing is timed to take place during slack water before the ebb so that the net can be more readily controlled; the changing times of low and high water are memorised by the fishermen in a chanted ditty (see also Arunachalam, 1996: 268, regarding similar memory training in Tamil Nadu). The *patia* is manhandled from its berth above high water down to the water's edge using bamboo poles slotted through rope grommets at three or more oar stations, and a large seine net is loaded into the stern sheets. This net is shaped like a funnel, the mesh size decreasing towards its bag end. To the wings of the net are attached lines which can be two kilometres in length. One or two men are left on the beach tending the end of one line and the boat is launched and pulled through the surf, bows-on to the breakers, a passage that is more tedious but less dangerous than the process of landing (Hall, 1868: 318).

Oared *patia* generally have eight oarsmen (*danri*), although one boat was seen with nine rowing stations. The bow oarsman sits on the foremost crossbeam and pulls an oar to starboard, the next oar aft is to port. The other six oarsmen are double-banked, two to a crossbeam (Fig. 3.19). As in the sailing *patia*, oars are pulled against a rope grommet attached to side timbers which protrude above the boat's sheerline. The steersman (*majhi*) stands to his task, using a steering oar which is longer than that used in the sailing *patia*. This may, in theory, be pivoted through one of two grommets, to port

Figure 3.19 View towards the bow of a Digha oared *patia*, showing the rowing arrangements.

and to starboard of the stern 'post'. In all the boats examined at Digha, however, the starboard grommet was not fitted and the hole where it could have been fastened appeared to be unworn: Digha steersmen evidently prefer to have their steering oar to port. In addition to the steersman and eight oarsmen there are two or three men to handle the net.

Once through the surf, the boat is steered in a large semicircle, normally not more than a kilometre from shore, and the net is payed out continuously. By this time the team manning the line has been increased to fifteen or more men who track the boat as she circles. When the boat reaches the shore, having passed through the surf once again, the end of the second line is passed to a team of similar size. The two teams begin to haul in the net, gradually moving towards one another. Although the boat may be used for only an hour or so, the hauling-in can take three or four hours; as with beach seining throughout the world, the results are frequently disappointing. Further details of seine-net fishing from the shore are given in pages 131–135.

Seine-net fishing was said to be undertaken at Digha mainly during the rainy season, March to August, although operations can be curtailed in June and July during the monsoon period. One boat was observed beach seining during November at Digha, and four were also seen at Talesari in the same month. As elsewhere on India's east coast, the crew work for an owner who does not go to sea. The steersman is 'in command' and, for example, calls the stroke for the passage through the surf, which is always hazardous, especially during landing. He may be the only permanent member of a crew.

The social context

Kalavathy and Tietze (1984: 13) recorded that, in Balasore District, 81 per cent of fishing households own one or more nets, but only 17 per cent own boats. It is probable that a significant number of non-fishing households own boats, but there is an expectation that every crew member of a boat will contribute a net in order to claim his share of the catch.

The Kaibarta caste is usually associated with fishing. However, at Kasafal, the fishermen belonged to more than one caste – indeed the crew of the boat measured were Harijans.

Whether or not fishing and other seafaring activities have ever been exclusive, the historical importance of the maritime wealth of Orissa is still reflected in folklore and folktales and commemorated in cultural festivals. The boat is featured in, and forms an integral part of, the festival of Kartika Purnima, when thousands of tiny boats of cork, coloured paper or the bark of the banyan tree (*Boita Bandana*) are set afloat to celebrate Orissa's maritime past (Patnaik 1982: 58; Patra 1988: 11). The festival of Chaiti Ghoda is also of great importance to Orissan fishing communities (Patnaik, 1982: 23–7). A culture in which boats find so special a place must have had a long association with the sea and, one assumes, a rich maritime past.

The *patia* and other South Asian reverse-clinker boats

Reverse-clinker planking is found both in India (discussed in this chapter) and in Bangladesh (see Chapter 2). The Bangladesh boats of Sylhet District differ from the *patia* of Orissa/West Bengal in a number of ways:

- Sylhet boats are used on rivers, rather than in estuaries and coastal waters.
- Sylhet boats have a wider variety of forms, including flat-bottomed as well as round-hulled.
- Sylhet boats have shallow-angled, block stem (*goloi*) or plank-*goloi* ends, rather than steeply rising, hog/false stem ends.
- The reverse-clinker planking of the *Sylheti nauka* is 'low rise' and ends on a (near) horizontal, full-length line, the central part of which is formed by a specially shaped, short, reverse-clinker strake; while the *patia*, towards the ends, has 'high-rise', strongly curved runs of planking terminating on diagonal lines. In this respect, the planking of Solvyns' late-eighteenth-century *pettoua/petooa* (Fig. 3.4) is more like that of a Sylhet boat than of a *patia*. This may suggest that the high-rising, reverse-clinker planking of the *patia* may be an innovation of the nineteenth, or even the twentieth, century.
- The methods of closing the ends differ. The *patia* method is complex, whereas the *nauka* solution is much simpler, as befits a river boat.
- The reverse-clinker planking of Sylhet boats (and of Solvyns' *pettoua/petooa*) is 'capped', and the upper hull completed, by full-length planking which consists of standing, reverse-clinker or European clinker strakes, or a combination of these; whereas the space between the two ends of the rising, reverse-clinker planking of the *patia* is filled by a standing strake and several European clinker strakes (Fig. 3.5).
- Boatbuilders' staples and spikes are used as plank fastenings in Sylhet, rather than the Orissan/Bengali nails clenched by hooking (see Fig. 2.3, lower).

In general terms, however, the Bangladesh and the Orissan/West Bengali boats are similar. The planking differences listed above are due, in the main, to the different operating environments which necessitate different shapes to the boats' ends. The seagoing *patia* has rising ends; the Sylhet river boats have low ends, the angle matching the general slope of river banks.

Why use reverse-clinker planking?

Two periods of fieldwork in northern Orissa and southern West Bengal have shown that the reverse-clinker technique is used to build *patia* with two distinct functions in two different environments: coastal and estuary fishing with drift nets; and beach seine netting through surf. That one boat type can undertake both functions, in two different coastal environments, suggests that environment and function are not determinants in the choice of the reverse-clinker boatbuilding technique. Indeed, its use in this region seems to be primarily a cultural rather than a technological choice. The fact that reverse-clinker planking is (and has been) also used to build boats in nearby Bangladesh (see Chapter 2) suggests that this technique was once much more widely used in the northeast of South Asia than it is today.

The problem of the interpretation of iconographic evidence for reverse clinker and the associated problem of 'hulc planking', and the impact of both these features on European medieval maritime studies, are considered in more detail in the Appendix.

4

THE SMOOTH-SKINNED TRADITIONAL INLAND BOATS OF BANGLADESH

Colin Palmer

This chapter is based on the author's practical field experiences as a consultant naval architect visiting Bangladesh on many occasions between 1985 and 1993. It draws on the results of more than two hundred interviews with boatmen from all over the country and on detailed studies of nine individual boats (BIWTA, 1994), but it is particularly concerned with the smooth-skinned boats that operate on the inland water and canals of Bangladesh – boats that Bengalis refer to as of *binekata* construction. In Chapter 2 we discussed the much less common *digekata* (reverse-clinker) boats, which these days appear to be almost exclusively restricted to the northeast of the country.

The term *binekata* refers to the method of construction in which the hull is constructed plank first and the planking is arranged so as to give a smooth surface: there are no visible overlaps between the planks. This method of construction provides considerable scope for variations in hull form, and many different hull shapes are indeed found within Bangladesh and in neighbouring West Bengal in India.

The boats

Names of boat types

Several authors, including Hornell (1946: 241–53), Greenhill (1971: 83–125), Rahman (1968: 5–7), Jansen (1989: Chapter 5) and Deloche (1994: 156–70), have commented on the large number of boat types in Bangladesh. The problem of the naming of these types is also discussed in Chapter 2 above, but no entirely satisfactory system has been proposed. There is general agreement that the generic term for the boats under discussion is *deshi nauka* – literally, 'country boats' – but the naming of individual types is very confusing. The heart of the problem is that, when describing or identifying a boat, local people and boatmen think in terms of location and type of operation just as much as of a particular boat shape. Consequently, two boats that appear identical to outside eyes may be called by different names by their respective operators or owners. Conversely, outwardly different boats may be given the same type name.

The study by the Bangladesh Inland Water Transport Authority (BIWTA, 1994) also recognised this difficulty of terminology and attempted to classify boats on the basis of external appearance as perceived by outsiders and the names most commonly given to boats of these shapes by boatmen. The project field studies (which were conducted in all parts of Bangladesh except the far northeast) recorded less than forty different boat names – significantly fewer than earlier authors such as Rahman (1968). It is not known whether this represents a recent reduction in the diversity of boat types. Within the range of names recorded, some were far more frequent than others. Such types as the *ghashi*, *soronga*, *dingi*, *panshi*, *patam*, *kosha*, *malar*, *raptani* and *bachari* were among the most common. By far the commonest was the *kosha* – (also known as 'shallow' because they have been fitted with diesel engines originally used to power shallow-tube well water pumps). Because the names of boats appear to refer to a generalised set of features, one name may embrace a wide range of sizes. Figure 4.1 shows the range of sizes for the main types of boats recorded in the BIWTA study: the names *kosha* and *soronga* can apply to boats which range in size from less than 100 *maunds* to more than 1000 *maunds*. (The *maund* is a measure of weight and there are approximately 27 *maunds* to one metric tonne.) Other names do not span quite such large size ranges, but all except the *chandi* and *dingi* apply to size ranges covering several hundred *maunds*.

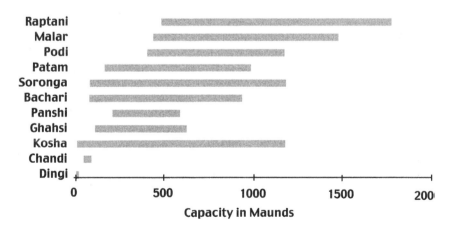

Figure 4.1 Size range of eleven boat types.

Names of parts

While the naming of boats may be complex and confusing, the boat builders and boatmen have a more consistent set of names for the parts of their boats. The *goloi*, discussed in Chapter 2 above, is a distinctive feature of the boats of Bangladesh, and there are comparable specific words in Bengali for many other parts of the structure.

The useful life of boats

Country boats made of high-quality wood can have a very long life. Several of the boats recorded in interviews conducted during the BIWTA study were said to be more than thirty years old, but these are exceptions that are becoming increasingly rare. The combination of the reduction in wood quality and the revival in boatbuilding resulting from mechanisation means that almost 40 per cent of boats are less than five years old and only 20 per cent are more than twenty years old. This distribution suggests that the boat population is changing quickly, as old boats disappear to be replaced by a greater number of new ones. Linked to the age distribution is the change in the types of boats. The old boats tend to be the traditional types which require large quantities of high-quality wood for their construction; new boats are much more likely to be the utilitarian and economical *kosha* type.

The shapes of country boats

At the most general level, country boats are either spoon-shaped (round-bilged) or flat-bottomed. There is considerably more diversity in the shapes of round-bilged boats than among the hard-chine *koshas*. This is probably because the former represent the long-term tradition and have had longer to change under the influences of generations of boatbuilders and owners. Greenhill (1971: 86) and Jansen (1989: 74) have referred to the presence or absence of a *goloi* as an important distinction in the classification of the different types. It is certainly an important identifying feature, but in terms of practical operation or hydrodynamic performance the *goloi* has little significance, although it is sometimes used to provide a convenient 'gangplank' from shore to boat.

Among the round-bilged types most commonly found, it is possible, from a technical standpoint, to classify them in terms of cross-section shape and heaviness of construction. On the basis of cross-section, the two extreme forms are the *patam* and the *podi*. The former is almost semicircular in cross-section – 'spoon-shaped' – whereas the latter is 'firm-bilged': it is like a rectangle with rounded corners. Between these extremes are types such as the *soronga* (close to the *patam* at the spoon-shaped end of the range), the *ghashi* and *kathami*, which are intermediary, and the *raptani*, which is almost as firm-bilged as the *podi*.

The *patam* and *podi* use relatively 'light' construction, whereas the *raptani* and *kathami* use 'heavier' construction. The words 'light' and 'heavy' are used here in the meaning they have in naval architecture: a heavily built vessel has relatively thick planks, large cross-section longitudinal members and close-spaced frames, as compared to a lightly built one.

Figure 4.2 has been compiled to show the relationship between cross-section shape and heaviness of construction for the selected boat types, including the flat-bottomed, hard-chine *kosha*, which represents the extreme development of the firm-bilged hull form.

It is impossible to identify all the historical influences which have resulted in the different hull shapes of Bangladesh country boats. Factors such as the type of cargo may influence the weight of construction (e.g. the *kathami* is heavily built, presumably to resist the loads imposed by the cargoes of irregularly shaped logs which it carries). However, the *kathami* is also built in the Barisal area, where wood is relatively plentiful and low cost, so the owners can perhaps afford to invest in a stronger boat in the expectation that

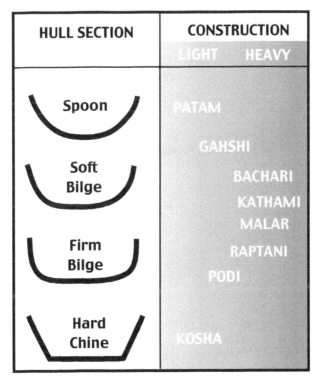

Figure 4.2 Relationship between hull shape and weight of construction.

it will have a longer life. The *patam* and *soronga* are both used to transport building materials, but the *patam* is more lightly built.

Typical boats

As noted above, several authors have commented on the recent history of the boats of Bangladesh, but few details of the construction of specific boats or their hull shapes have been published. This paper therefore sets out to make a step in that direction rather than simply attempting another synthesis of existing published material. The BIWTA study provided a unique opportunity to obtain detailed information on selected boats. In all, the project worked on and recorded nine different boat types: *chandi*; *ghashi*; *soronga*; *podi*; *patam*; *kosha* (two types); *raptani*; and *malar*. The project also had access to a substantial photographic record as well as a measured plan of the *bachari* type. A subset of three, the *malar*, *bachari* and *patam*, has been selected from the nine types to represent the range of hull forms that can be found – from the firm-bilged *malar* to the extreme spoon shape of the *patam* (Fig. 4.3).

The *malar* is to many eyes (particularly those of their proud and prosperous owners) the finest of all the boats. *Malars* carry a large press of sail (Fig. 4.4) but are also rapidly accepting the benefits of mechanisation. Most are based in Noakali District and for this reason they are sometimes called

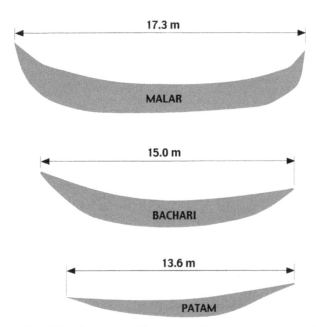

Figure 4.3 Profiles of the three selected boat types, drawn to a common scale.

102

Figure 4.4 A *malar* under full sail. Photo: Reidar Kvam.

noakali boats. *Malar* are general-purpose bulk cargo boats, which often carry jute or salt. In the past they were built to carry as much as 3,000 *maund* (120 tonnes), but 1,000 *maund* (40 tonnes) is now more typical.

The *bachari* (Fig. 4.5) is from the Khulna area and has a hull form that is distinctly different from that of the *malar*. The *bachari* hull is much more rounded and lies between the *malar* and *patam*. *Bachari* are some of the more heavily built country boats – in part perhaps because of the ready availability of wood in the Khulna area but also because of the trade they traditionally

Figure 4.5 A *bachari* dried out on the riverbank in readiness for retarring the bottom. Note the internal structural arrangements.

ply, carrying logs from the Sundarban forest which is in the delta region to the south of Khulna (see Fig. 2.1).

The *patam* (Fig. 4.6) has the most rounded, spoon-shaped hull of all and is also one of the most lightly built boats. This appears to conflict with the work the boat often performs – carrying boulders and building materials from the northeast to Dhaka.

These three types are used as examples in the sections that follow and Table 4.1 records typical dimensions of some of their timbers (scantlings – see Glossary). Figures 4.7 to 4.9 record the overall dimensions of these types and their hull lines.

Figure 4.6 Light and loaded *patam* sailing in a gentle breeze. Note how the boat sailing close hauled (centre of picture) is also being rowed from the bow. The *patam* is unusual in having a bow higher than the stern.

Table 4.1 Typical hull scantlings of selected boats

| Member | Dimension *mm* | Boat type | | |
		Bachari	*Malar*	*Patam*
Planking	Thickness	32	32	32
	Width	200	200–250	225
Floor timbers	Siding	75	57	75
	Moulding	165	150	190
	Spacing	400	380	380
Bilge futtocks	Siding	67	64	n/a
	Moulding	120	64	n/a
	Spacing	400	380	n/a
Side futtocks	Siding	67	64	75
	Moulding	120	64	57
	Spacing	400	380	380
Crossbeams	Width	108	90	75
	Depth	125	100	115
	Spacing	580	1650	760
(upper) Stringers	Width	45	27	38
	Depth	125	170	100
(lower) Stringers	Width	45	27	n/a
	Depth	250	190	n/a

The boatbuilders

Boatbuilding starts with the owner collecting capital from various sources to provide at least 50 per cent of the cost. To do this, the boatman may call on family sources – perhaps the sale of land or cattle, or a mortgage over other land. These sources are preferred to moneylenders, who charge as much as 10 per cent interest per month. The balance of the cost is provided by way of credit from traders for the purchase of wood, staples etc. and by deferring payments to boatbuilders. The amount of credit that can be secured depends upon the status and wealth of the boat owner. Such credit arrangements are generally informal and verbal, not written down.

Historically there was an extensive network of boat owners, wood traders and boatbuilders. The boatbuilders had strong links with both the wood suppliers and the boat owners and played something of an intermediary role in the arrangements. The boatbuilders were artisans who enjoyed respect, so the boat owners had to maintain friendly relationships as any problems with the builders meant delay in the progress of boat construction. There was also strong community solidarity among the boatbuilders, which put pressure on

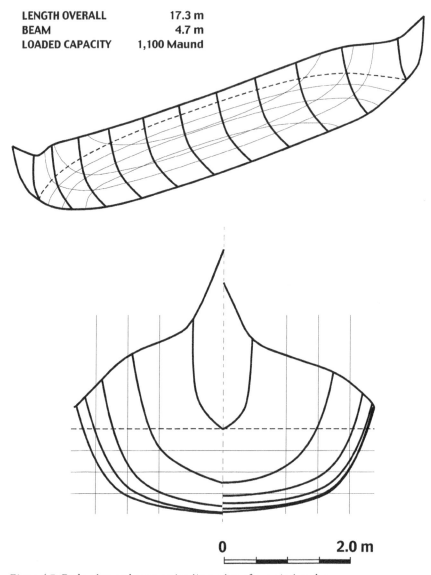

LENGTH OVERALL   17.3 m
BEAM   4.7 m
LOADED CAPACITY   1,100 Maund

0   2.0 m

Figure 4.7 Body plan and perspective lines plan of a typical *malar*.

the owner to cooperate with them. In recent times these established patterns have come under pressure as a supply of high-quality wood has become increasingly difficult to obtain owing to the felling of forests. This has reduced demand for large, skill-intensive boats. Smaller, simpler boats are taking their place, and these can be built by house carpenters who now compete with traditional builders.

106

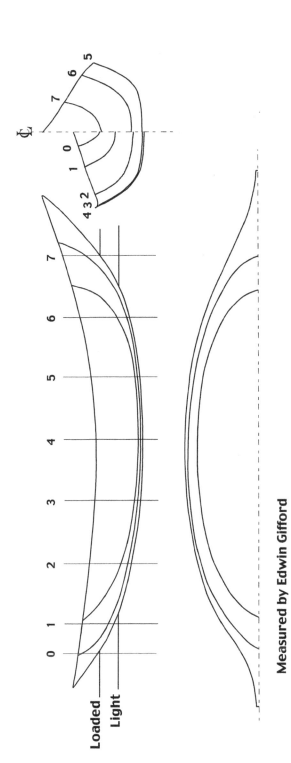

Measured by Edwin Gifford

Overall Length 15.0 m
Beam 4.42 m
Capacity 700 Maund

Figure 4.8 Lines plan of a *bachari* measured at Khulna in 1980, as redrawn from Howe, 1981.

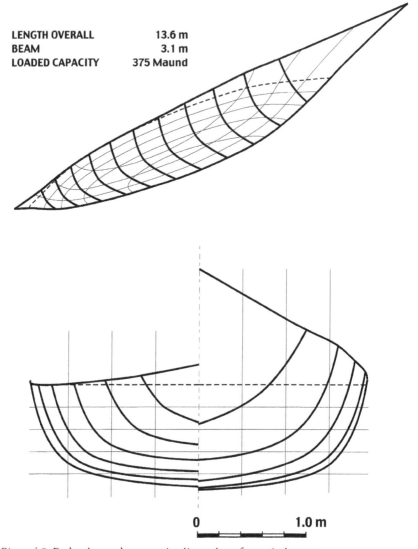

LENGTH OVERALL	**13.6 m**
BEAM	**3.1 m**
LOADED CAPACITY	**375 Maund**

0   **1.0 m**

Figure 4.9 Body plan and perspective lines plan of a typical *patam*.

Construction method

Chapter 2 deals specifically with Bangladesh reverse-clinker boats and describes the plank-first approach and the use of thin steel staples (called *patam* – the same word as the name of the boat type) to fasten the edges of the planks together (Fig. 4.10). This edge-joining technique is used on all the boats described in this section, but in such a way as to give a smooth hull

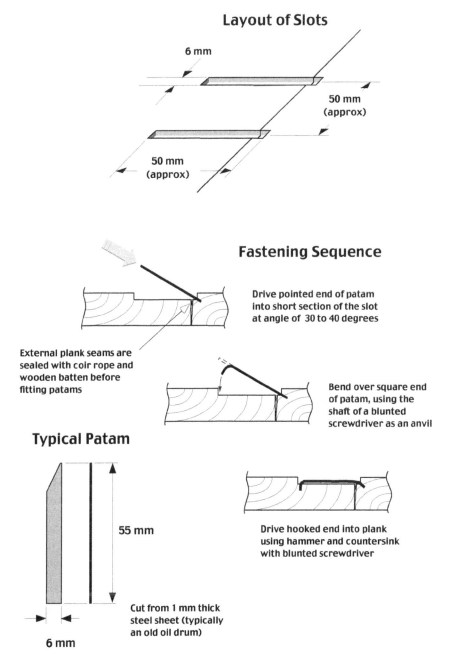

Layout of Slots

6 mm

50 mm (approx)

50 mm (approx)

Fastening Sequence

Drive pointed end of patam into short section of the slot at angle of 30 to 40 degrees

External plank seams are sealed with coir rope and wooden batten before fitting patams

Bend over square end of patam, using the shaft of a blunted screwdriver as an anvil

Typical Patam

55 mm

Cut from 1 mm thick steel sheet (typically an old oil drum)

6 mm

Drive hooked end into plank using hammer and countersink with blunted screwdriver

Figure 4.10 Illustration showing the layout of the edge joining staples (*patam*) and their fitting sequence.

surface (*binekata*) rather than the stepped surface of the reverse-clinker boats (*digekata*).

The choice of wood used for boatbuilding depends on factors such as geographical location as well as price. Thirty or forty years ago, *sal* and *sundari* were widely used for boatbuilding. Currently *sal* is hardly used, and only rich boatmen can afford *sundari*. *Garjan* is now commonly used in the Chittagong area for building inland as well as offshore boats. Other species used include *sil koroi*, *chambul*, jackfruit, *rendi koroi*, blackberry and mango. Bamboo is the most common material for the superstructures and platforms.

The descriptions that follow are based on many observations of the repair and construction of boats in Bangladesh, during the period from 1988 to 1993. Because the new construction of traditional boats is now rare, it was not possible during any visit to watch a traditional type of boat being built from start to finish. Just one opportunity to witness the complete construction of a hull was possible, but, in that case, the boat was a launch (motor boat) built on a keel-stem backbone. However, despite this introduction of what might be termed a 'Western' feature, the hull was still built plank first, and no frames whatsoever were fitted before the planking reached gunwale height. Although the boat was to have a transom, this was not fitted until all the planking was complete.

It appears therefore that the techniques used by the boatbuilders (*mistris*) have become 'hybridised' and that, if there ever was one single method, it has now been adapted and modified to cope with changing times. The account that follows does not therefore describe the construction of just one boat but is an amalgam of information from different observations that will serve to give an impression of the approach taken by the builders of today.

One feature that is found on all these boats is the use of what is termed 'hulc' planking in Chapter 2 and in the Appendix. This planking pattern is very widespread in the boats of Bangladesh, but it is not used exclusively. In the descriptions that follow, the variety of hulc planking arrangements that exist on the *malar*, *bachari* and *patam* will be described and compared.

The principles of the construction of all three types are broadly similar. The boats are built on an open area of riverbank, generally during the dry season (November to March). The foundation plank (central bottom plank) is laid first and set in the curve that defines the longitudinal curvature of the hull. It is not clear exactly how this is done, but it probably entails a combination of supporting blocks and heat bending. The bottom planking is edge-joined to the foundation plank with staples (*patam*) and sweeps up at bow and stern to be terminated in the 'hulc' style. Greenhill (1971: 76) has described how, 40 to 50 years ago, the planks were fitted with a form of lap join and this has been quoted by more recent authors (e.g. Jansen and Bolstad, 1992: 13). However, there is some evidence from photographs and personal observation that this technique of cutting lap joints on the plank edges may not be universal. More study is now required to ascertain if the lap join is used selectively or if it is becoming obsolete.

The plank faces are coated with river mud and the planks then heat-softened over a smouldering fire of wood shavings so that they can be bent to shape using a system of levering. Where twisting is required, one end is fastened in its position on the hull and the other twisted by means of open-mouthed levers that are tied down to stakes in the ground. Once the planks have cooled and the shape is set, they are trimmed to fit and the twisting/ bending process is repeated until they can be positioned with the normal combination of levers and manpower.

Once in position, the planks are given the necessary 'dishing' or transverse curvature by the system of levers shown in Figure 4.11. These levers are only applied while the plank is being positioned and fastened to its neighbour with the internal row of staples. Once fastened, the plank is propped from the outside, if required, and the levers are removed. At the bow and stern the planking may require substantial bending and/or twisting. Twisting is less likely with the traditionally shaped boats, but in the case of launch types it can be considerable at the bow. The hull is planked up to and around the turn of the bilge before any transverse floors or framing members are fitted. Indeed in some cases the planking is more or less completed before any transverse members are fitted. The timing and sequence of fitting the frames appear to be a matter of personal preference for different boatbuilders.

The first transverse stiffening members to be fitted are the floors (*baka*) (as shown, for example, in photographs in Jansen 1989: 82). They are fastened with long metal (forged steel) spikes called *gazal*: these are made from scrap steel, hand forged in a range of sizes. Where outside access is difficult, some spikes are driven from inboard (N. Rahman, pers. comm.), but the great majority are driven from outboard through the planking and floors, and their points are bent downwards (turned through 90 degrees) onto the inner face of the floors. Floors are followed by the futtocks. Not all boats have intermediary bilge futtocks fitted, but it is not clear whether this choice is determined by size, shape or type of boat, or by a combination of these factors. Floors and futtocks seldom, if ever, touch each other and they are never joined together. The continuity of the hull structure is ensured by the overlap between the ends of the floors and futtocks. Typically, this overlap is three to four plank widths. The precise location of the futtocks is varied so as to cover plank joins wherever possible, but this is not a practice that is followed exclusively. Where plank butts within strakes are required, a special joint is used: this is illustrated in Figure 4.12.

The spacing of the framing sections is regular over most of the length and is typically around 0.40 m between centres. In the midbody of the boats the futtocks are arranged to lie on a plane that is at right angles to the centreline

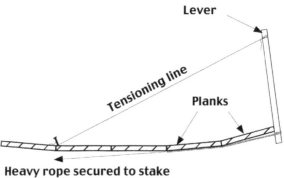

Figure 4.11 The planks are positioned against each other using a system of levers which apply pressure to the planks and enable the builder to angle the planks relative to one another.

Figure 4.12 A distinctive hooked joint is used to join plank ends together.

(see Fig. 4.5). However, at the ends, where there is a rapid narrowing and increased curvature, the floors continue in the same pattern, but the futtocks become more irregular and are canted so that they do not have to be highly bevelled (Fig. 4.13).

Drilling and shaping

The boatbuilders have a complete set of hand tools for all the drilling and shaping work they need to do, but if power tools are available they use them. Planks are cut from a log using mechanical band saws wherever possible. However, when there is no power, a hand sawing system is still used, exactly as described by Greenhill (1971: 73).

When planks need to be shaped during the construction process, the tool most commonly used is the side axe, or occasionally an adze. For other shaping work the boatbuilders also use chisels which commonly have a wide blade and are bevelled on both faces. However, the builders have also adopted what might be called Western-style chisels bevelled only on one side. For drilling holes a bow drill is most commonly used. However, in the southwest of the country other types of drills were observed, for example hand-made wooden pump drills.

Figure 4.13 Interior structure of a *malar*, showing how the futtocks are canted in the ends of the hull. Note also the hulc planking runs and the way in which the fastenings are bent over onto the surface of the futtocks and floors.

Transverse and longitudinal structural members

Once planking is complete and framing installed, the hull is finished by fitting longitudinal stringers (*derenga*) and transverse beams (*gura*). Stringers are not fitted on all boats, notably the *patam* (but the top plank on the *patam* is trapezoidal in section and thus provides additional structural strength). A distinctive feature is the way that the longitudinal members (stringers) are joined (Fig. 4.14). The longitudinal members are fastened to the upper frames by metal spikes.

Transverse beams are positioned to lie beside a frame section, but they do not rest upon a longitudinal member (as might be expected). The beams are notched into the planking and project a short distance outboard of the hull. There is no local strengthening at the ends of the beams.

Strake patterns

As noted earlier, the construction of a boat typically commences with the laying of a central plank, to which pairs of planks are fastened on either side to create the hull. A distinctive feature that results (and is found on all three described types of boat) is that the ends of most planks do not terminate on a

Figure 4.14 The same style of hooked joint as in Fig. 4.12, used to join stringers or inwales.

stem or stern post as in familiar Western-style boatbuilding. Instead they run upward, more or less parallel to the line of the stem or stern, and terminate beneath one or more longitudinal planks: this pattern, as noted above, is known as 'hulc' planking. Figures 4.15 to 4.17 illustrate this pattern on a *patam*, *bachari* and *malar*. In all these figures, the plank seams have been emphasised by overlaid lines wherever it was possible to distinguish them on the original photographs.

Figure 4.15 shows the stern and bow of *patams* (two different boats, but very similar vessels, nonetheless). The pattern of the strakes is clear at the stern and shows the 'hulc' system of termination beneath a single sheer strake. This process of terminating strakes under the sheer strake is discontinued part way along the hull: at this point, more or less horizontal planks are fitted, the ends of which terminate on the side of a 'rising' plank.

Figure 4.16 shows the plank runs at the bow and stern of a *bachari*. They terminate under the sheer strake, but the discontinuity identified on the *patam* is not present. Instead, in the midbody, the two planks immediately beneath the sheer strake are given considerable shaping to provide the longitudinal curve that suits the run of the lower planks. The plank immediately beneath the sheer strake is different from the rest in that its ends are sharply tapered to allow the plank beneath it to terminate on the sheer strake.

Finally, Figure 4.17 shows the stern of a *malar*. The three upper strakes run the full length of the hull; the lowest of these provides the face for the termination of the lower strakes. The *malar*, like the *bachari*, has one plank that tapers to a point to allow this pattern to blend smoothly.

Figure 4.15 Stern (upper) and bow of *patam* with the strake runs enhanced for clarity.

Strake breadths

It has been suggested (Greenhill, 2000: 15) that one characteristic of hulc planking is that it permits constant-breadth planks to be used. No measurements of plank breadth of actual boats are available, so, to investigate this suggestion further, Figure 4.18 was prepared. This is a bow view of a

116

Figure 4.16 Strake runs at the bow and stern of a *bachari*.

Figure 4.17 Strake runs at the stern of a *malar.*

bachari photographed on a slipway in Zinzira, the boatbuilding area of Dhaka. The actual strake runs on the starboard side (left side of the photograph) have been highlighted, as in previous photographs. The overlay on the port side, on the other hand, is derived from a lines plan prepared for the boat. Using a process of measuring along the section lines, strake runs of constant-breadth planks have been marked in.

It is immediately apparent that the appearance of the strake pattern developed from the lines plan is different from that of the actual *bachari*. The curvature of the seams is outboard on the 'constant width' lines plan (right side of Fig. 4.18), but inboard on the actual vessel (left side of Fig. 4.18). This suggests that the planks on the *bachari* are not of constant width but are tapered towards bow and stern. Confirmation of this suggestion will require measurements to be taken on actual boats, as it is possible that parallax or perspective errors may be influencing the conclusions drawn from the photograph.

Conclusion

Binekata-built boats represent a distinct tradition that was once widespread on the rivers and inland waters of what are now Bangladesh and West Bengal. The traditional methods of *binekata* construction, and the boats it

118

Figure 4.18 Bow view of a *bachari* showing actual strake runs (left) and calculated strake runs for planks of constant breadth (right).

produces, are fast disappearing under pressure from modernisation and the scarcity of the wood required. Fortunately, the boats of Bangladesh have been studied by a number of authors over the last 150 years, so in comparison with many other boatbuilding traditions, the information available provides a relatively comprehensive, if not complete, record. *Binekata* boats are notable for their diversity of form, ranging from extremely long and narrow fishing and racing boats to firm-bilged, heavily built cargo carriers like the *malar* and *raptani*.

While this chapter and the other sources mentioned provide a broad picture of this diversity, as well as many details of the construction, naming and operation of the boats, a few important areas remain where further study is important before the techniques and knowledge are lost forever. For example, it is unclear how the initial setting-up stages of construction are planned and executed; how the floors are fastened in the early stages of construction; why the extent and type of framing vary so much from type to type; how widely the distinctive plank lap is used; and the extent to which parallel-sided planking is used.

5

THE *MASULA* – A SEWN PLANK SURF BOAT OF INDIA'S EASTERN COAST

Eric Kentley

Of all sewn plank boats, the *masula* is possibly the one best represented by models in museum collections (no museum is known to have a full-size example) and it is frequently illustrated in general works on ethnographic craft. Yet only sparse and often contradictory accounts of it have been published: so much so that Prins (1986) was able to classify it only tentatively. This chapter attempts to provide a detailed account of the *masula*'s form, construction and use. This, however, is not an account of a single boat but a range of boats. Although *masulas* share certain common features, they are not of a single design. The materials used in construction differ from one stretch of the coastline to another. There is also some diversity in usage, in method of propulsion and even in method of sewing.

The original data was collected during two fieldwork periods on the coasts of eastern India (Fig. 5.1): 6 January–16 March 1983 and 9 January–9 February 1984. During the first period the area between Nagappattinam (10°46′N) and Bimlipatnam (17°57′N), just north of Visakhapatnam, was surveyed, during the second the area between Santopilly (18°01′N) and Paradeep (20°16′N). The method was to move from town to town along the coast, rapidly visiting all the readily accessible fishing villages, returning to those with particularly significant boats or informative builders or fishermen. It was not possible to visit every village and it was difficult to find boats under construction, and so much of this report is based on interviews. Variations may therefore exist that are not recorded here.

Shorter versions of this study have been published in Kentley (1985), Kentley (1993) and Kentley (1996).

Historical and modern sources on the *masula*

Early accounts by Europeans of sewn boats on India's east coast refer to these craft by a variety of related labels: *massola, massoolah, mossel, mussoola, macule* etc. (Hill, 1958: 207). In the twentieth century the name has become

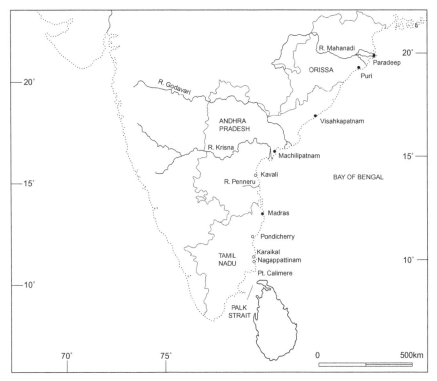

Figure 5.1 Map showing sites on India's eastern coast.

largely standardised as *'masula'*. However, the origin of the term is not known. It is not a word currently employed by, or even known to, the builders and users of this type of craft. Although Hill (1958: 208) dismisses the theory, it may be derived from the town Masulipatnam (which was translated for me as 'Muslim town'), now renamed Machilipatnam ('Fish town'): this is a theory widely accepted among fisheries department officials throughout the coast.

Hornell (1920: 175) states that these craft are known to the fishermen of the Coromandel coast (which can be defined as the coastline from Point Calimere to the mouth of the River Krishna) as *padagu* or *salangu*. I have not heard the latter term in use, nor had it recognised. *Padagu* is the word used by Tamil speakers, *padava* is the Telugu term in Andhra Pradesh. In Orissa, it is pronounced more like *padua*. However, none of these terms is used exclusively for sewn boats: some metal-fastened boats are also so labelled. Therefore, as *masula* is used specifically to mean a sewn boat of this region, albeit not by its operators, and has a widespread currency among those interested in ethnographic craft, there is a good case for its retention and I will continue to use it here.

121

Perhaps the oldest surviving sketch of a *masula* (Fig. 5.2) was drawn by Thomas Bowrey, who was in India during the latter half of the seventeenth century:

> The boats they doe lade and unlade ships or vessels with are built very sleight haveing no timbers in them save thafts to hold their sides together. Their planke are very broad and thinne, sowed together with cayre, being flatt bottommed and every way much deformed ... They are so sleightly built for conveniences sake and really are most proper for this Coast; for, all along the shore, the sea runneth high and breaketh, to which they doe buckle and alsoe to the ground when they strike. They are called Massoolas, and are for little use save carryinge of light goods (as bailes of callicoes or silkes, not exceeding 6 or 8 at one time).
>
> (Temple, 1905: 42–3)

Witsen (1690 – cited by Prins, 1985) describes the craft as 'useful boats but have to be taken to parts and resewn often'. Dr Fryer, another seventeenth-century visitor, notes that, in addition to sewing the planks with coir rope, the boats are 'caulked with dammer (a tree gum or resin)' (Fryer, 1698). He also states that the boats are steered by two men aft 'using their paddles instead of a rudder', as Bowrey's sketch suggests.

The handling of the *masula* is well described by Gaspardo Baldi in a sixteenth-century account (quoted in Hill, 1958: 208):

> ... merchandise and passengers are transported from shipboard to the town by certain boats which are sewn with fine cords; and when they approach the beach, where the sea breaks with great violence, they wait till the perilous wave has passed and then, in the interval between one wave and the next, those boatmen pull with great force and so run ashore; and being there overtaken by the waves they are carried still

Figure 5.2 Sketch of a *masula* by Thomas Bowrey (seventeenth century).

further up the beach. And the boats do not break, because they give to the wave; and because the beach is covered with sand the boats stand upright on their bottoms.

By the nineteenth century this ship-to-shore service, in Madras at least, was organised by the Government, with log rafts in attendance in case of mishaps. Edye (1834: 8) gives further details of the sewing of the *masulas*: 'The planks which form these boats are sewed together with coir yarn, crossing the seams over a wadding of coir, which presses on the joints and prevents leakage'. He also notes that the boats are guided by two steersmen and provides a drawing. This is less informative than Bowrey's sketch. No indication is given of how the thwarts or crossbeams (Bowrey's 'thafts') are fitted. Bowrey at least indicates that they project through the hull at the upper edge of the strake below the top strake; Edye does not represent them at all. Although in his description Edye does not state whether the coir wadding is placed over the seams inboard or outboard or both, his drawing and Bowrey's indicate that it is at least outboard, the sewing itself forming a zigzag pattern, except on Edye's drawing at the stem and stern posts where it is shown as unconnected bars of stitching. However, engravings by T. and W. Daniell (1797) and aquatints by Hunt after East (*c.* 1837) display a different pattern: unconnected bars are shown outboard on the strake-to-strake seams, with no wadding visible except on the upper part of the stem post. Furthermore, the boats are shown as being guided by a single steersman.

Admiral F.E. Pâris (1843) provides the best detailed construction drawings of a *masula* (which he calls a 'chelingue' – a term, like *masula*, which is not recognised by the users of the boat), which differs in several respects from those considered above (Fig. 5.3). On the side strakes the wadding is depicted outboard, held in place by a type of cross-stitch consisting of 'vertical' bars connected by pairs of diagonals. Inboard, the pattern is depicted as double unconnected bars. On the bottom strakes this is reversed: the wadding and cross-stitches are inboard and the vertical bars outboard. Additionally, unlike Bowrey, Pâris shows the crossbeams passing through the top strake. However, he also shows the stem and stern posts extending below the plank bottom of the boat, a feature perhaps suggested by Bowrey, but not by Edye.

Folkard (1870: 309) gives a different account of the materials used in construction:

> ... the planks and other parts are sewn and laced together with the strong fibres of the cocoa-nut tree, layers of cotton being placed between the planks. Over the seams, inside, a flat narrow strip of tough fibrous wood is laid, the whole then being jointed to stout stem and stern posts in the same manner.

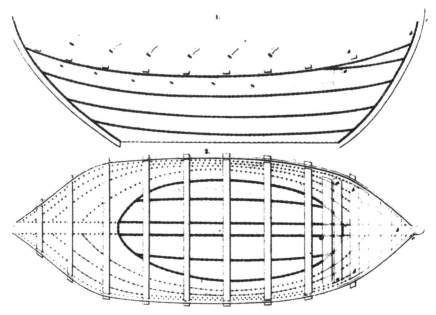

Figure 5.3 Masula by F.E. Pâris (1843).

Thus he contradicts Pâris's drawing and earlier statements that the wadding material is coir. He is also the only writer to suggest that any material is placed *between* the planks. His sketch depicts the sewing pattern in the same manner as the Hunt aquatint and shows the crossbeams penetrating the hull in the same position as on Bowrey's drawing.

Models from Madras of *masulas* held by the National Maritime Museum, Greenwich, and dating from the nineteenth century accord fairly well with the drawings of Folkard and Hunt. Outboard the sewing on the strake-to-strake seams forms a pattern of unconnected bars; inboard it mainly forms the same cross-stitch pattern as depicted by Pâris, over a wadding material (which is not a wooden batten). At the stem and stern posts this criss-cross pattern is again found over the wadding material, both inboard and outboard. The stem and stern posts extend below the plank bottom: a feature, as I have suggested, which is indicated in the drawings of Bowrey and Pâris. As on the drawings of Bowrey and Folkard, but not of Pâris, the crossbeams on the models are shown as penetrating the hull at the seam of the top strake and the strake below.

Twentieth-century literature seems to support the general accuracy of these models and to cast doubt on some of the earlier drawings, such as those by Edye. James Hornell, perhaps the greatest authority on ethnographic craft to date, who was in the early years of the last century Director of Fisheries in Madras, states categorically that Edye's drawing is inaccurate:

The stitching of the planks is wrongly shown as a zig-zag line of twine over a 'wadding of coir' on the outside. As a matter of fact this wadding is on the inside of the boat. Looking at the outside the lacing is shown solely by vertical bars of cord between two parallel rows of holes along each side of the jointing of the planks.

(Hornell, 1926: 58).

By implication, the drawings of Bowrey and Pâris also stand condemned: provided of course one accepts Hornell's implicit assumption that in a hundred years a change has not taken place. On the other hand, although Hornell may be confirming Folkard's sketch, Folkard's description is not endorsed, as Hornell states that the wadding is of coir, not a wooden batten.

Hornell also noted that *masulas* were no longer in use as passenger carriers and that only in the smaller ports of the Coromandel coast were they employed as cargo carriers. However, many *masulas* 'of a smaller size than those for cargo work are used in shooting shore seines during the fine weather' (Hornell, 1926: 58). Elsewhere more information is given on the boat:

Its range extends along the whole of the East Coast northwards of Cape Calimere. On the Coromandel Coast it is distinctly short in proportion to its beam and depth ... In the Vizagapatam and Godavari districts it often runs longer and more elegant ... As the masula boat never goes far from shore, mast and sail are not used; a crew varying from 8 to 12 in number perched insecurely on thwarts far forward provide the motive power with long paddle sweeps not less than 12 feet in length; steering is done with a very long and powerful paddle suspended in a coir grommet from the projecting head of the stern post. No iron is used in the hull ... A narrow keel projecting about 2 inches is normally seen, no bulkheads are used, stem and stern are both raked considerably and both are somewhat bluff in their curves. Freeboard has to be very high as they have to pass through heavy breakers and in consequence the loads they carry are light compared to their apparent capacity. An oculus is sometimes (rarely) painted on the bows at Madras. In the seining masulas working in the neighbourhood of Uppada, Vizagapatam district, this type of boat attains its greatest development, both in regard to size and economic importance. There it is esteemed the most valuable asset of fishermen and these men have rigorously developed the use of the shore seine and without roomy boats would be unable to carry and handle the great length of net requisite to effective operations on a large scale. These Vizagapatam (Visakhapatnam) fishing masulas are usually decorated along the sides with two rows of sloped panels of alternating black and white, those of the upper row slanting opposite to the inclination of the lower ones.

(Hornell, 1920: 174–5).

Hornell's sketch of an Uppada *masula* shows three strakes which do not meet the stem and stern posts – that is, three stealers – whereas both Pâris's drawing and museum models show only one stealer.

In *Water Transport*, Hornell published a photograph of a Madras *masula* (Hornell, 1946, plate 1). This seems to be a boat under construction or reconstruction. The crossbeams are positioned on the top strake, but the sewing holes are visible on the top edge and on the stem post, so it is certain that another strake is still to be added. The wadding material is plainly to be seen outboard on the stem post-to-strake seams, and the strands at the top appear to be a type of grass, not coir. The pattern of sewing on this seam is unfortunately not clear. There are certainly bars of sewing as on the strake-to-strake seams, but there may also be a cross-stitch (which in fact Hornell draws on another sketch [Hornell, 1923: 291] consistent with Folkard and Hunt).

Zeiner and Rasmussen (1958: 27), in a general survey of Indian fishing boats, base their section about the *masula* almost entirely on Hornell (1920) but state that the planking is caulked with plantain-leaf stalks, which De Kerchove (1961: 499) also mentions. Zeiner and Rasmussen give a figure of 1,400 *masulas* on the Madras coast (Tamil Nadu) and 1,000 on the Andhra Pradesh coast in 1957, each costing between Rs 1,000/- and Rs 3,000/-. Blake (1969: 58) puts the number of *masulas* on the Madras coast alone at 5,000. He adds that the woods used in construction are *asnamaram* (Indian laurel, *Terminalia tormentosa*) and mango (*Mangifera indica*).

Suryanarayana's study (1977), like Blake's, is sociological in focus and deals only briefly with the craft of the fishermen, but he makes two important observations: firstly, in Andhra Pradesh not all fishing villages have *masulas* and, secondly, two sizes of *masulas* are built in Visakhapatnam – the *pedda* (big) *padava* and the *chinna* (small) *padava* (see Table 5.1), costing Rs 400/- and Rs 250/- respectively. The larger takes two men seven days to construct, the smaller five days. The manner of construction for both sizes is identical: '[the boat is] prepared out of mango planks without any ribs or frames. The planks are sewn together with coir rope and the interspaces between them filled with dry straw' (Suryanarayana, 1977: 26).

'Interspaces' presumably means the space between the rows of sewing holes on two planks when they are butted together. As it is not stated whether the interspaces are filled inboard or outboard, it may be implied that they are filled on both surfaces. This would be borne out by a measured drawing of a Visakhapatnam boat by Gulbrandsen (1979). He identifies the dry straw mentioned by Suryanarayana as *rella gaddi* (*Sacchurum spontanium* or *Vetiveria zizandiodes*) and the planking as a combination of mango and *sirsini* (*Ficus glomerata*). It should be noted that the transverse members on this boat do not penetrate the hull and are relatively much broader than those shown by Pâris and on the models. No stealer is indicated, nor is there a decked area at the stern. The sewing pattern is indicated by a rough zigzag line, not entirely dissimilar to that found on Bowrey's sketch.

Table 5.1 Published dimensions of *masulas*

Length, breadth, depth in metres	L	B	D	L/B	B/D
Edye (1834) gives values between	10.67	3.35	2.44	3.19	1.37
	9.14	3.04	2.13	3.01	1.43
Pâris (1841) measured from drawing	9.90	3.40	2.00	2.91	1.70
Hornell (1920): Visakhapatnam	12.19				
Pondicherry	8.53	2.44	1.22	3.50	2.00
De Kerchove (1961) gives values between	10.67	3.35	2.13	3.19	1.57
	7.93	2.44	1.22	3.25	2.00
Suryanarayana (1977)					
pedda padava	9.14		1.52		
chinna padava	6.71		1.22		
Gulbrandsen (1979)	7.95	1.37	0.62	5.80	2.21
Mohapatra (1983)	7.42	2.20	1.14	3.37	1.93

Mohapatra (1983: 16–17) does not include a drawing but states that the *masulas* of Orissa are built (by implication entirely) of mango. He gives the basic dimensions (see Table 5.1), *including those of its mast and sail*. No other writer indicates that the *masula* is propelled by any means other than oars.

Table 5.1 lists the dimensions of *masulas* where they are given in the literature considered. The earlier values describe a craft that is deep (following McKee's definition [1983: 79] of deep as B/D $\leqslant$ 2.00), but Gulbrandsen's drawing is of a narrow boat (L/B $\geqslant$ 3.75), a very different-looking craft from those illustrated by Bowrey, Edye, Hunt, Pâris, Folkard and Hornell.

All accounts agree that the *masula* is a frameless sewn boat. But, as has been seen, they give diverse information on many other aspects of the craft: its sewing pattern, the wadding materials used and their position, and the fitting of the crossbeams. One writer even gives an alternative means of propulsion. It is possible that each account is accurate, the authors having observed the boats at different times and in some cases on different stretches of the coastline. Furthermore, Bowrey, Edye, Hunt, Pâris and Folkard portray a cargo- and passenger-carrying vessel, while Hornell, Blake, Suryanarayana, Gulbrandsen and Mohapatra observed fishing vessels. Over a period of 400 years, it is more than likely that some changes in design and construction have taken place.

However, although Hornell (1920) is the first writer to mention fishing *masulas*, it cannot be concluded that the omission of this observation in earlier accounts means that the fishing *masula* has been developed, relatively recently, from the cargo/passenger carrier. Fishing *masulas* may have always existed alongside the carrier and may even pre-date the carriers.

Distribution and use

From the sources discussed in the preceding section we can derive only a very general definition of the *masula*. It is a frameless plank boat, which is sewn with coir rope and is found on India's eastern coast. It is in fact the only sewn plank boat on this coast (but see Chapter 7 for the *vattai*, which has a sewn top strake).

In 1983/84 the distribution of *masulas* could be summarised as follows (see Fig. 5.1): the most northerly point where they were found was Paradeep, the port of Orissa; the most southerly was Chachapadi, a fishing village in Tamil Nadu about 11 km north of Karaikal. Between these two points there was not a continuous distribution. There were none between Paradeep and Puri and also none between Dainapeta, in Andhra Pradesh, about 100 km south of Visakhapatnam, and the coastal villages in the vicinity of Kavali – not even in Uppada where Hornell (1926) made his sketch, nor in the perhaps eponymous Machilipatnam.

There appeared to be just one *masula* between the River Pennar in Andhra Pradesh and Madras (at Pulicat – an 'import' from further south). In the remaining areas, from Puri to Dainaipeta, from Kavali to the River Pennar and from Madras to Chachapadi most, but not all, fishing villages had *masulas*.

The boundaries of this distribution are constantly changing and the areas in which *masulas* were found were even then contracting. An informant told me that he had seen them at the mouth of the River Chandhrabhaga, north of Puri, in 1980. Fishermen in Uppada said that they had not used *masulas* since the 1960s, when they had abandoned beach seining.

With the exception of a very few used as ferries connecting fishing villages across river mouths or backwaters, *masulas* are now exclusively fishing vessels. They are worked by Tamil and Telugu fishermen. Although the boats are found in Orissa, they are used by immigrants or, more commonly, by descendants of immigrants from Andhra Pradesh – that is, Telugu speakers. I have been told that there are a few Oriya fishermen working alongside Telugu, but I have not encountered any. Paradeep is the most northerly point of the distribution of Telugu immigrants, but the distribution of *masulas* does not entirely coincide with this. Telugu fishermen are found north of Puri, but only in Paradeep are there *masulas*.

Some explanation of the boat's distribution lies in its function and the environment in which it must operate. The majority of *masulas* perform just one type of fishing operation: shooting a beach seining net. This is carried out where the underwater slope of the beach is relatively steep. For example, at Ennore, near Madras, where the United Nations Food and Agriculture Organisation's Development of Small Scale Fisheries in the Bay of Bengal programme conducted trials of beach landing craft, the slope is 20 degrees (Gulbrandsen *et al.*, 1980: 2). *Masulas* are not found where the continental shelf broadens south of Nagapattinam in Tamil Nadu nor north of Puri (except at Paradeep, where the *masulas* are not used as beach seiners). Nor are they found around the river mouths of the large rivers, most notably the Krishna-Godavari region, where the build-up of silt results in a gentle underwater slope.

Beach seining is undertaken much further north, in Digha and Talesari, but with much smaller nets and boats on low-declivity beaches (see pages 93–95). However, within the area of the *masula*'s distribution there are a few places where beach seining is conducted by other craft. For example at Satankuppam and other villages in the vicinity of Pulicat, rivetted plank boats are used. I am told that, in Andhra Pradesh, the metal-fastened *nava*, used primarily for deepwater handlining, can also be used for shooting seine nets. In Karailkuppam, there was no *masula*, but there was one beach seining net. I was told that this was shot by a seven-log raft.

Economic and social factors also play a role in the distribution and numbers of *masulas*. Madras fishermen stated that in the 1960s catches were of ten tons per haul; by the 1980s they were down to one or two tons. Nowhere on the coast did I even see this amount being landed. Usual catches seemed to be only a couple of baskets – about 50 kg – fetching Rs 200/-. There was said to have been an 80 per cent reduction in the number of *masulas* in Madras from 1960 to the mid-1980s and the trend appears to have continued into the 1990s. The blame is laid on the mechanised trawlers, introduced in 1955. Although they do not compete for the same catches – the *masulas* catch pelagic shoal fish, the trawlers prawns – the trawlers are said to disturb the shoals. The trawlers have other effects: they are perceived as a good investment for prosperous fishermen who might otherwise invest in *masulas* (or more *masulas*); and they also take labour away from the fishing communities which might otherwise have been employed in the beach seining operation, which requires substantial manpower. It was also suggested to me that another cause of labour shortage has been the introduction of free school meals in Tamil Nadu and Andhra Pradesh, which has taken away boys who previously would have helped to haul in the nets.

Although fishermen and fisheries department officials agree that the *masula* is in decline, I have been unable to find historical records detailing numbers and distribution with which to compare the present situation. The 1979 census of beach landing craft is reproduced in Table 5.2. In addition to these there were at this time about 2,000 mechanised boats on the whole coast.

Table 5.2 shows clearly that the most common craft are log rafts, of which there are two basic types – the *teppa* of the northern area and the *kattumaram* of the southern. However, bearing in mind the geographical distribution of *masulas* described above, it may seem strange that Andhra Pradesh should have more of these boats than the other two regions combined, especially as it has fewer log rafts than Tamil Nadu and Pondicherry. Table 5.3 shows that it cannot be attributed to a larger fishing population.

Table 5.2 Estimates of the numbers of traditional beach landing craft in 1979

	West Bengal & Orissa	Andhra Pradesh	Tamil Nadu & Pondicherry	Total
Log raft	4,300	18,100	26,200	48,600
Masula	300	4,100	3,000	4,700
Nava	500	1,100	–	1,600
Other craft	1,400	2,500	5,600	9,500
Total	6,500	25,800	32,100	64,400

Source: Gulbrandsen, 1979

Table 5.3 Fishing population of India's east coast

	West Bengal & Orissa	Andhra Pradesh	Tamil Nadu & Pondicherry	Total
Active fishermen	15,076	64,592	72,102	151,700
Total fishing population (including women and children)	61,082	237,470	305,000	603,552

Source: Gulbrandsen, 1979, based on 1973–7 Census

Obviously, the numbers of craft recorded in Table 5.2 are rounded. They may be an underestimation. According to figures obtained from the Orissa State Fisheries Department there were 25 *masulas* at Paradeep, 232 in the area between the southern tip of Lake Chilka and the Andhra Pradesh border, and none in the area in between. Yet numerous *masulas* were seen in Puri. But this was in January: had researchers visited at a different time, in June for example, they would not have been so fortunate. In June there are no *masulas* in Puri. They are all dismantled and stored away.

Such an error could not have occurred in Tamil Nadu, where the boats are not dismantled after the season, but it could be committed in Andhra Pradesh, where some are dismantled. This would mean an underestimation of a figure that already seems strangely high. The explanation is that in Andhra Pradesh, as Suryanarayana (1977) notes, there are two sizes of *masula*: the larger a beach seiner, the smaller used primarily for setting gill nets. The latter, which will be discussed more fully below, was developed as a substitute for the *teppa* log raft and in Andhra Pradesh is more numerous than the beach seiners. These small *masulas* were exported to Paradeep, where they worked out of the harbour, setting nets mainly in the area between the harbour and the mouth of the River Mohandi – an area where, because of a high sea wall, it is impossible to land. There are no beach seiners here, and these small *masulas* are found nowhere else in Orissa, not everywhere in Andhra Pradesh and not at all in Tamil Nadu.

Beach seining

The small *masula* is used all year round, weather permitting, but beach seining is a seasonal activity, restricted to three or four months of the year. These are normally December to March in Orissa and January to April in Tamil Nadu. This season, which brings the pelagic shoals of anchovies, sardines, whitebait, silver bellies and caranx in towards the land, is the latter part of the northeastern monsoon period. Seining is normally performed in the early morning and, if successful, further seines may be shot until about midday. It is rarely performed in the late afternoon. The seining net is known in Telugu as *pedda vala* and in Tamil as *peria valai* – literally, 'big net', a fair description. The 1980s' price was up to Rs 15,000/-, more than the boats themselves.

The net (Fig. 5.4) consists of four parts: the cod end, the main body, the wings and the towing warps. Mohapatra (1983) gives the following account of the net in Orissa. The cod end comprises an upper and lower half which when folded out form a rectangle composed of 88 pieces – 8 across and 22 lengthwise. Each piece is 50 meshes across by 110 lengthwise, the mesh size (the distance between knots when the net is stretched) varying from 10 mm at the extreme end to 15 mm where the net joins the main body. The main body itself is 12.34 m in length and is trapezoid when laid flat, the mesh size increasing from 17 to 62 mm where it meets the wings. At this point it is 9 m deep. The wings are composed of three parts. The first, of cotton, is 32 m long with a mesh size of 90 mm; the middle portion, of coir, is 54.86 m long with a mesh size of 900 mm; and the rope end, of hemp or synthetic fibre, is 137.16 m long with again a mesh size of 900 mm.

The float and sinker lines are 10 mm diameter coir ropes along the wings and 7 mm cotton on the main body. On the float line, 80 wooden floats (of *Erythrina sp.*), each about 600–700 mm by 100–150 mm, are attached at approximately half-metre intervals. On the sinker line, stone sinkers of $\frac{1}{2}$ kg to 1 kg in weight are fastened opposite each float.

The towing warps are composed of 30 on one side of the net and 50 on the other. Each rope is 22.5 m in length, 37 mm in diameter. Two hemp or

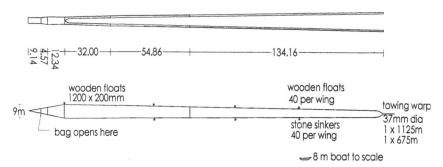

Figure 5.4 Dimensions and features of a beach seining net.

synthetic ropes and one coir rope, or alternatively one hemp and one coir rope, are joined in series until the correct number of ropes is made up.

There are many variations in size. In Tamil Nadu, the cod end is trapezoid when laid flat. One measured was 9.10 m along the extreme end, 4.00 m at the main body, with sides of 8.60 m. The main body was said to be 8.80 m long, the wings 440 m, the shorter warp 30 m and the longer 728 m. The warps are normally composed of only coir ropes. Hornell (1924C: 74) gives the following dimensions of a net (he does not state where he measured it): cod end – 7.28 m in length with a width of 6.07 m at the extreme end and 3.03 m at the main body; main body – 8.49 m with a width forward of 18.20 m. He divides the wings into two parts: the wings proper – 15.17 m; and the leaders – 303.3 m. The warps are 909.9 m; he does not note any difference in size between the two ends.

The net is loaded into the *masula*, forward of the decked area upon which the steersman stands, leaving the end of the shorter warp in the care of one or two men on the beach. However, in some places where there is a lot of beach seining activity, it may be necessary for a fisherman to swim to the shore with the end of the warp once the boat has been rowed into a suitable position.

With only the steersman and two or three oarsmen on board, the boat is pushed into the water bow first by the crew (which can number between eight and fifteen in total). Once it is afloat, the crew will jump in and take up their positions. The boat must then be rowed through the heavy surf that beats on all the beaches where the seiners operate. This surf is the result of waves breaking far out in the Bay of Bengal. The underwater slope of the beach governs the rate at which the energy of the waves is dissipated. On most beaches where the *masulas* work, the surf zone is relatively narrow and the wave energy is rapidly released from a wave moving at high speed, which becomes a plunging breaker as it nears the shoreline. On very steep beaches a plunging breaker will break with great force on the face of the beach (Gulbrandsen *et al.*, 1980: 2).

The surf is normally higher in the area between Kakinada (16°57′N, 82°14′E) and Puri (19°48′N, 85°49′E) than elsewhere. Measurements taken by the Meteorological Observatory between Kakinada and Calingapatnam (18°21′N, 84°08′E) from 1950 to 1955 show only 15 per cent of days with wave heights of less than 0.75 m, with usual heights up to 1.8 m (Gulbrandsen *et al.*, 1980: 2). Wave periods are generally between seven and ten seconds.

Currents normally run parallel to the coast at speeds of $1\frac{1}{2}$ to 4 knots. The strength and direction of these do not correspond with the monsoons: the northward flow is strongest in November, the southward in March (Hydrographer of the Navy, 1978: 52).

The steersman will know the local wave pattern, which consists of a number of small waves and a following single or multiple big wave (Gurtner,

1960). Once the big wave has broken, the boat will be rapidly rowed out so that the bow will be well over the next wave's crest as it breaks. The boat will then slam down with breathtaking force. If the bow is not over the crest, the boat will be pushed back and possibly swamped. However, a steersman will be sufficiently experienced to know if the speed he is achieving is insufficient and will abandon the attempt well beforehand, allowing the boat to be pushed back to await the next opportunity.

The boat is rowed out eastwards, or slightly to the southeast if there is a northward-flowing current, slightly to the northeast if the current is southward. One or two men who do not row shoot the net. If the current is northward, the warp is let out to port. Once the wing of the net is reached, it is put out to starboard (the steersman must duck!). Sometimes, once the cod end is released the boat will turn northwards, shooting the rest of the net and bringing the end of the other warp back to land. On other occasions, the crew will continue to row until the end of the second wing is shot before turning. Either way, the boats only operate within 3 km of the shore.

Thus the net is laid out so that it crosses the current and encircles the shoals that tend to swim against the current. The boat will then return to land about 250 m north of the shorter warp. If the current is southward-flowing, directions will be reversed: the boat will circle to the south of the warp.

The steersman standing on the small stern deck is able to locate the breaking point of the inrolling waves. He will aim for the boat to be brought in by the big wave in the pattern. Gurtner (1960: 593) describes the correct procedure for beaching boats through surf. The boat's centre of mass must be kept slightly behind the wave crest at breaking point, so that the boat will ride in like a surf board. If the centre of mass is forward of the wave crest, it will be overrun by the breaking wave and may be swamped or broach.

The hauling-in of the net commences immediately. The boat crew are involved, with help from as many other males as are available on the beach. There may be up to fifteen men on each side of the net. The longest warp is hauled first until both warps are reckoned to be equal length. Then both are hauled, the two groups of men gradually moving towards each other. A common technique is for each man to have a short length of coir rope around his waist and looped loosely over the warp. He will then walk backwards until a spot is reached where one man will be in charge of coiling the rope. As the net is brought in, it is disconnected from the ropes and laid out to dry. Shooting the net takes about half an hour; hauling it in may take two hours. In some areas a small, very simple one-man raft may be launched to guide in the cod end, but this is not common. As the cod end nears the shore (Fig. 5.5), two or three men will plunge into the water, disconnect it from the main body and drag it back to land.

By this time, women will have arrived and, once the fish have been shaken out of the net, they basket up the fish and take them to market or to the

Figure 5.5 Final stages of hauling in a seine net at Madras: men are in the water to disconnect the cod end.

village for drying. If the catch has been good, the boat may be launched again. If not, it will be hauled up the beach and over a slight crest, beyond the range of the water.

The boatbuilders

As noted earlier, the builders and users of the *masulas* are either Telugu or Tamil people. The majority are Hindu, although there are some Muslims, particularly in the Madras area. The builders are not a separate caste, but men living within fishing communities and fishermen themselves. Normally they are taught the skill by relatives. It is frequently the case that the grandfather of a builder will have taught him – the skill skips a generation. Not every fishing village has a boatbuilder, and a builder will often travel to neighbouring villages to construct boats. An exception is Thota Veedhi, a quarter of Bimlipatnam, Andhra Pradesh, where a large number of builders live and work, exporting their products by cart or by sea to the villages to the north and the south. A similar community may exist at Bandersam, near Calingapatnam, but I have been unable to confirm this.

The builders of Bimlipatnam do not construct rafts, although elsewhere all boatbuilders appear to be able to do this. None make domestic furniture or build houses commercially.

Masulas are constructed in villages, not on beaches, although repairs are effected there. The villages offer shade, security for tools and closeness of sustenance. The wood for construction is supplied by the prospective owner, who is always a fisherman. Although some villages have saw pits, these are now abandoned and the timber for strakes is cut in saw mills, where the logs are sawn through-and-through (i.e. tangentially to the growth rings). Only one builder is required for construction and he will do all the woodwork, but he will be assisted by a number of men (normally not less than three) who will do most of the sewing on the boat. The technique of sewing is widely known among fishermen, and most builders rely on labour from the community where the boat is being built. Some builders however have a team of sewers who travel with them from one construction to the next. If the work is continuous, it is reckoned to take one builder and three sewers seven days to build a beach seining *masula*.

Except at Bimlipatnam, where small *masulas* are produced, little new construction now seems to be taking place. Most building is simply the replacement of worn strakes (particularly the garboards) and in some areas the seasonal resewing of boats (in October or November).

I have not seen the commencement of constructing a new boat, but the procedure appears to be as follows. An auspicious day is chosen for laying the plank-keel and the whole village assembles. Flowers, *betel* and plantains are laid on the plank as an offering and cooked rice is distributed. Incense sticks are burned and prayers said, while a coconut is broken over the plank.

Once the construction is complete, the boat is dragged to the shore. On the sheer strake near the bow, or on the stem post, a smear of milk, sandalwood paste and turmeric is made, in the centre of which a *taluk* (the red dot Hindu women put on their forehead) is placed. Those present take a *taluk* from the same dish. The owner has plantains, *betel* and sweets distributed. The builder will then take a small chip of wood from the smeared area and give it to the owner, who will place it among his household deities. Finally, a coconut is broken over the bow and the boat will be launched for a symbolic (rather than a fish-catching) journey.

Types of *masula*

The *masulas'* most distinctive features are their sewn planking and the absence of frames. They are also all fine-sterned. In general, they are simply designed boats: the shape is created by the manipulation of the width of the bottom planking, the breadth of the boat at the sheerline, and the rake of the stem and stern posts. However, there are differences in construction techniques in size (even among the beach seiners), in decoration and, even though all beach seiners are rowed, in rowing arrangements. With such variety and such limited historical data, no particular example of a *masula* can be taken as a 'prototype' of which all others are variations.

However, the differences between *masulas* are not random, except to some degree in decoration. The most significant differences are the method of sewing (of which there are two), the form of the second strake, the form of the stem or stern post/keel-plank joint and how the boat is rowed. These variations can be tabulated to reveal three principal types of *masula*. These three types are not found throughout the area of the east coast suitable for beach seining but are each confined to a specific geographical area, almost coinciding with the state boundaries of Orissa, Andhra Pradesh and Tamil Nadu. As they do not coincide exactly, the terms northern, central and southern sector are used in preference. A tabulation of the differences is shown in Table 5.4. Each of the three types will now be examined in detail.

Table 5.4 Principal differences between *masulas*

	Principal sewing method	Form of second strake	Post/keel plank connection	Oarsmen to each oar
Northern-sector *masula*	Method 1	Stealer	Half-lap, keel plank lap below post lap	Can be more than one
Central-sector *masula*	Method 1	Tapering plank	Half-lap, keel plank lap below post lap	Can be more than one
Southern-sector *masula*	Method 2	Stealer	Half-lap, keel plank lap above post lap	One only

Northern-sector masulas

The northern-sector type of *masula* is found from Puri to the River Vamsadhara in Andhra Pradesh (Fig. 5.6). A few of these boats are used as ferries crossing river mouths and backwaters (for example at Navalarevu and Gopalpur). All others are beach seiners.

This *masula* is a five-strake boat (Figs 5.7 and 5.8), although strictly the second strake is not a strake but a large stealer. It has six or seven crossbeams and a decked area at the stern. Most boats of this sector are between 7.5 and 8 m in length, although (with the exception of one ferry) the smallest measured was 6.75 m and the largest 8.40 m. Length-to-beam ratios are around 3.5 and beam-to-depth ratios about 2.

The central longitudinal member is normally of *Shorea robusta* (*sal*, which is the main planking wood of the metal-fastened boats of northern Orissa), but Indian laurel or even, allegedly, teak (*Tectona grandis*) may be used. The stem and stern posts are of *Syzigium cuminii*. All planks are mango, including those that make up the stern deck (which are often made from old strakes). The crossbeams are of casuarina (*Cashuarina equisetifolia*). The boats are sewn together with two-ply coir rope, with the dried marsh grass *rella gaddi* as the wadding material on most seams.

The central longitudinal member on boats of this type is a plank-keel: that is, its breadth is greater than its depth, a handspan (about 16 cm) wide and a hand's width (8–12 cm) in depth, rectangular in cross-section. It should be 8 cubits in length (a cubit being the length from the elbow to the

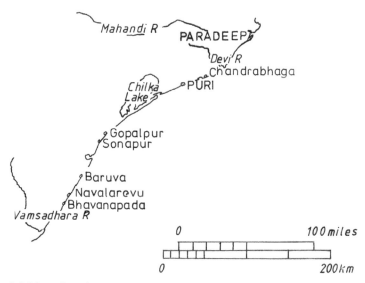

Figure 5.6 Map of northern sector.

137

Figure 5.7 Masula at Puri.

fingertips). The ends are slightly tapered and are fastened to the stem and stern posts by horizontal half-lap joints, the projecting lap of the keel plank being below that of the post. The length of the lap is calculated as the width of six fingers and is secured by sewing over a wadding of coir fibre (*not* dried grass) inboard.

The stem and stern posts are adzed into shape from logs, normally tapering from the same breadth as the plank-keel to about 8 cm at the sheerline. Rarely is either a single piece: both posts are usually made of two parts, again scarfed by a horizontal half-lap joint. The lap on the lower section is inboard. The two parts are sometimes nailed together but are more commonly held in position by the sewing that binds the strake ends to the posts. There is a consensus that the stem post should be 7 cubits in length, measured on the curve, but 5 and 6 cubits are the values given by different builders to the stern post.

Flat on their inner aspect, the surface of the posts which takes the ends of the strakes is at 90 degrees to this. The outer surface is usually gently rounded, except in the southerly parts of this sector, where it is more wedge-like. The stem-post head retains the shape, but the stern-post head is commonly adzed almost to a point, so that the rope against which the steering oar works can be slipped over easily and made firm. Alternatively, the stern-post head may be indented on the starboard side or on both sides.

All planks are of the same thickness, 22 mm – seven threads (a thread being an eighth of an inch). Strakes are not necessarily one-piece but are rarely more than two. To make up a strake, the planks are often simply butted together, although they are sometimes plain scarfed and sewn together. (However, at

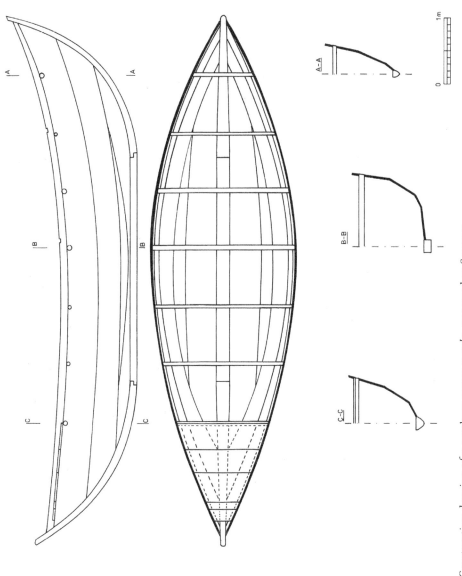

Figure 5.8 Construction drawing of a northern-sector *masula*, measured at Sonapur.

Bhavanapada one boat was found which had its fourth strake on the port side composed of two planks which were connected by a wooden piece [butt strap] nailed on the inside.) Coir is frequently used as the wadding material on such joints, inboard and outboard, whereas dried grass is used on the strake-to-strake seams. It was not possible to obtain an explanation for this. Certainly, coir does not rot as rapidly as grass and is regarded as a superior wadding material. However, it is much more expensive than grass. But the use of grass as the primary wadding material means that the boats must be dismantled after every fishing season and the grass replaced, although this has the advantage that the boats are not left on the beaches during the cyclone season.

The garboard strake, 12 cubits long, butts against the side of the plank-keel and posts and is held in position by sewing over a wad of dried grass both inboard and outboard. It is positioned so that it extends beyond the plank-keel and onto the stem post, a distance of $2\frac{1}{2}$ cubits, and onto the stern post, thus forming the basis for the greater sheer forward than aft. For most of its length the garboard strake is parallel-sided, but it curves at the ends. This curve is roughly cut to shape and then trimmed *in situ* when the third strake is matched up, as the curve is determined by the third strake as it bends towards the posts.

Similarly the second strake, a stealer 8 cubits in length, has its gently curved upper side finally trimmed when the third strake is matched up. It ends where the curve of the garboard begins. Amidships, the second strake gives a total breadth from the edge of the keel plank to its upper side of $1\frac{1}{2}$ cubits. However in Edumanipelam, near Baruva, one boat was observed which had its starboard garboard and second strake combined in a single wide strake.

In Gopalpur there was a ferry which had only one strake between the second strake and the crossbeams. This boat measured 6.20 m in length with a maximum beam of 1.77 m and a depth of 0.42 m. It had five crossbeams and three old planks serving as thwarts for passengers. Otherwise it was structurally identical to the beach seiners. It was said to have been built entirely from old parts of a beach seiner and cost Rs 1,000/-, whereas the larger boats were costing Rs 8,000/-.

Again in Gopalpur, there was one beach-seining *masula* which had a single broad strake on both sides between the second strake and the crossbeams. Two other boats here had the same arrangement forward, but two planks aft. Normally there are two full strakes between the second strake and the crossbeams, approximately 16 and 18 cubits in length.

Halfway between Bhavanapada and Sonapily, a fishing village to the south, a *masula* was found with a frame for carrying the net. The top of the third strake had been indented to take three crossbeams. These were clamped into position by the fourth strake and onto them were lashed, longitudinally, 13 thin casuarina poles, each about 2 m long. This arrangement was seen nowhere else. This boat was also unique in having a drainage hole drilled through the keel plank about 15 cm from its start in the bow.

From the top of the fourth strake to the bottom of the third should be a distance of 1 m. If it falls short, a series of planks, a handspan or more in breadth, will be sewn onto the inner surface of the fourth strake, overlapping and rising above it to achieve the required height. Even if the fourth strake does give the required height, some builders will still fit this series of planks (not strictly an inwale, as it is not fitted to the top strake – it is convenient to call it an inner rail), but so that its top edge is flush with the fourth strake's top edge. This is said to give additional strength. Other builders say that it is unnecessary and merely adds weight. On some boats, particularly in Puri, this rail is fitted so that it rises above the fourth strake at the stern, thus raising the level of the stern deck, but flush with the top edge of the fourth strake along the rest of the boats. The technique for sewing on this inner rail will be described below.

The shape of the boat is held by the insertion of the crossbeams. As noted, most boats have six of these, but some have seven. They are roughly circular in cross-section and the fourth strake and/or inner rail (if present) is indented to take them. Rarely however do they lie flush with the top edge of the plank: normally they are slightly higher. They are lashed down with coir rope, holes being drilled in the fourth strake and sometimes through the beams themselves to take the rope. There is no standard method.

The fifth beam from the bow on a six-beam boat and the sixth beam on a seven-beam boat are not lashed down. They are removable, to aid loading the net. The first three crossbeams may protrude some distance outboard so that their ends can be used to lift the boat, but this is not common. Most beams are cut off roughly flush with the outer surface of the hull.

Measured along the curve of the fourth strake, the deck at the stern should extend 4 cubits. The planks that make it up are laid either on the top edge of the fourth strake or on the inner rail. Holes are drilled through them about 2 cm from the edge and they are secured by the sewing which connects the top strake to the fourth strake or inner rail.

The top strake is about 20 cubits in length and relatively narrow, between 80 and 110 mm. It is indented on its upper edge to take two thole pins on either side (propulsion is discussed in pages 159–161). Its lower edge will rest on the tops of the crossbeams, and if these have not been fitted flush with the fourth strake or the inner rail there will be a small gap. The planks which make up the top strake are not joined over a wadding of coir. They may in fact not be sewn together at all, save by the sewing that connects them to the fourth strake or the inner rail.

Sewing patterns

Sewing throughout the boat is with a doubled coir rope and on all strake-to-strake seams, except the top strake seam, it is done over a wadding of dried grass, both inboard and outboard. Builders are able to make a line of holes, three fingers apart and two fingers' width from the plank edge, by eye. The holes are bored with a bow drill, with a 1 cm bit. The lower plank is bored first, then the upper strake is matched up and marks are made so that each hole on this will be vertically above one on the lower strake. No material or treenail (a wooden fastening of circular cross-section – see Glossary) is fitted between the strakes. At certain points, particularly where the garboards join the posts, it is common to find that the dried grass has been covered inboard by a waterproof material, such as bits of plastic bags or bicycle-tyre inner tubes. The holes are plugged with balls of coir pushed in with a metal punch from inboard. However, on most boats the seam of the third and fourth strakes and the holes on the posts above this line are not plugged. It is thought unnecessary as it is above the waterline.

The top strake seam is never plugged. The pairs of holes on this seam are a handspan apart, and instead of dried grass the sewing is done over a 3 cm diameter rope (usually of hemp) inboard, and over numerous strands of coir rope outboard. The pattern of sewing on this seam is different in appearance to that on the lower seams: it is in fact 'half-sewn'. The sewing operation in this sector was not directly observed, but verbal descriptions indicate that it is identical to the practice found in the central sector. It will be described fully in pages 147–151. Briefly, the full operation consists of taking a coir rope from a point roughly amidships to the stem or stern post and then returning it to the starting point. On the top strake seam, the rope is taken in one direction only. This produces a criss-cross pattern on one side of the hull and a pattern of vertical bars connected by diagonals all running the same direction on the other.

The inner rail, when present, is sewn on using a different technique. As on the top strake seam, the pairs of holes are about a handspan apart. Where the rail is fitted flush with the fourth strake, the pairs of holes are bored through both planks; where the rail is fitted above, the lower holes go through both planks, but the upper holes are bored through the rail above the level of the strake. The coir rope is passed through a lower hole from inboard, then taken to the hole immediately above, passed through this and back down to the lower hole. It then travels outboard back up to and through the hole above and is taken (now inboard) diagonally to the lower hole of the adjacent pair. This sequence is repeated until the post is reached, where the sewing is finished off, as is all sewing, by knocking in a small wooden plug. Thus the pattern produced inboard is identical in appearance to that on the top strake seam, but outboard the pattern is of unconnected vertical bars.

Central-sector masulas

South of the River Vamsadhara, at Calingapatnam (Fig. 5.9), the design of the *masula* changes abruptly. From here to Dainaipeta one finds a six-strake

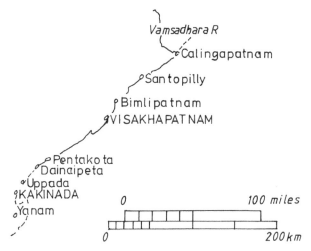

Figure 5.9 Map of central sector.

boat (Figs 5.10 and 5.11), probably the type that Hornell (1920: 75) described as the greatest development of the *masula*.

The major production centre for these boats appears to be Bimlipatnam, where, as noted earlier, there is a community of builders. They supply the villages from Kaya, 25 km north of Santopilly, to Dainaipeta. Further north, however, there are builders within the fishing communities, and in some relatively minor details their boats differ from those made in Bimlipatnam.

In Calingapatnam, the materials used in construction are the same as those used in the northern-sector boats, except that the fifth strake should be

Figure 5.10 Central-sector *masula* at Visakhapatnam.

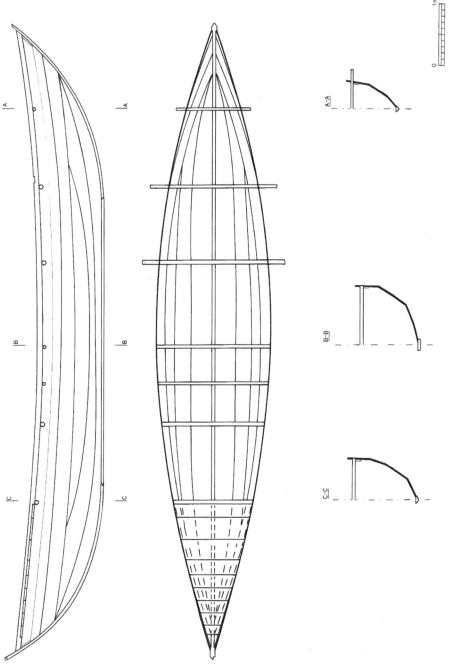

Figure 5.11 Construction drawing of a central-sector *masula*, measured at Visakhapatnam. Note the hulc planking pattern in the lower hull.

of Indian laurel. The Bimlipatnam builders may use mango for the posts and *Ficus glomerata* for the bottom three strakes. The remaining strakes may be mango, tamarind (*Tamarindus indica*), *Ficus religiosa* or banyan (*Ficus bengalensis*), depending on relative costs and availability. Everywhere *rella gaddi* is used as the principal wadding material, with coir for the plank-keel post joints and usually the plank-to-plank seams on a strake. The women of the Bimlipatnam community make coir rope by hand; elsewhere it is bought from small-scale manufacturers.

The beach seiners here are larger and narrower than those of the northern sector. In Calingapatnam they are over 8.5 m with a beam-to-length ratio of over 4. Those from Bimlipatnam are larger still, over 10 m, with a beam-to-length ratio of over 5. However, there are more differences than simply size between this type and that of the northern sector.

The plank-keel, 10 cubits in length, is fastened to the posts in the same manner as on the northern-sector boats. However the posts are more wedge-like in cross-section and do not taper. Again, they are normally in two parts but are usually nailed together (although they may be lashed by boring holes outboard and tying with rope). Characteristic of this type is the stern-post head, which is invariably indented on the starboard side to hold the steering rope. Instead of the garboard simply butting the side of a rectangular keel plank, a greater rise of floor is achieved by one of two methods. The Calingapatnam practice is to bevel the edge of the keel plank to create an angled surface for the garboard to abut. In Bimlipatnam, the keel plank is left rectangular, and the edge of the garboard is bevelled and stands on the top surface of the keel plank. The posts therefore differ in shape: in Calingapatnam there is a worked surface on the outer sides to take the ends of the strakes (like the northern-sector practice); in Bimlipatnam the strake ends butt against the inner surface of the posts. This is the theory. In practice one finds boats at Calingapatnam (and in the northern sector) the strake ends of which are fastened to the inner face of the stem post or stern post, or both. This is the result of accidentally cutting the strakes too short.

The garboard is positioned so that it extends an equal distance onto the stem and stern posts. For most of its length it is parallel-sided, but the outer edge begins to taper approximately at the ends of the keel.

The second strake should taper to meet the posts as a point. However, on some boats it does not quite reach one (or more rarely either) of the posts, and on others it reaches the posts with a little width left.

It is the third strake on these boats that is the stealer, not the second as on the northern-sector *masulas*. The fourth and fifth strakes complete the planking up to the crossbeams. On some boats the seam between these two strakes is wadded outboard with the midribs of palmyra leaves. An inner rail was present on all the examples examined. In Bimlipatnam this is fitted flush with the top of the fifth strake and is nailed on. In Calingapatnam it rises above the level of the strake and is sewn on. The method of sewing on this rail differs from the northern-sector practice and will be described below.

The fifth strake and/or the rail are recessed to take the crossbeams. There are seven of these, positioned $2\frac{1}{2}$ cubits apart, measuring along the curve of the fifth strake, except the fifth from the bow, which is placed half this distance between the fourth and sixth to give additional strength amidships. All are lashed down.

Some Bimlipatnam boats may also have one or more plank ties running transversely under the crossbeams. Two holes are drilled on the fifth strake through which a rope is threaded and tied off to a rope passing through the other side. One example was seen where the rope was tensioned by a form of Spanish windlass, with the bar jammed against a crossbeam.

The top strake, as on the northern-sector boats, is relatively narrow, recessed to take the thole pins and is 'half-sewn', with a rope inboard and numerous strands of coir rope outboard in the place of wadding.

No convincing explanation for the abrupt change of design on the southern side of the River Vamsadhara can be provided. There appear to be no political, cultural or economic differences between the areas north and south of the river. Those few fishermen and builders who know of the difference tend to ascribe it to that bane of anthropologists – 'tradition'. One builder in Calingapatnam stated that his grandfather had built boats of the northern-sector type but later changed to the central-sector type when this became known to him. A fisherman in Bhavanapada simply said that the northern-sector boats were suitable for their waters and those of the central sector more suitable for the waters upon which they operate, but he did not define what the differences in the waters were. It has been noted that the surf is heavier on the central-sector stretch of the coastline than elsewhere, but it seems unlikely that this alone accounts for the substantially different design in this area.

Masulas *not used for beach seining*

Beach seiners are not the only sewn boats found in the central sector. From Calingapatnam to Dainaipeta the small fishing *masula* – the *chinna padava* – is also used. This appears to be a fairly recent development. The earliest reference found is in Suryanarayana (1977). This type has emerged as a substitute for the *teppa*, the sailing raft of this (and the northern) sector, the wood for which (*Albizzia stipulata* or *Eythryna indica*) has become increasingly scarce and, consequently, expensive. A large *teppa* will cost approximately Rs 7,000/-, compared with Rs 2,000/- for the small *masula*. The small *masulas* found in Paradeep are brought from Bimlipatnam.

The boats are a little over 7 m long, so 'small' is a relative term: they are larger than some of the beach seiners of the northern sector. An example measured at Visakhapatnam measured 7.10 m in length with a length/beam ratio of 5.04 and a beam/depth ratio of 2.31. It is similar in design to the beach seiners of this sector, with the characteristic arrangement of the lower three strakes. However these boats have only four strakes per side with a narrow rubbing strake fitted flush with the top edge of the fourth strake and a double inner rail, again flush with the top edge of the fourth strake. Instead of the top strake being indented to take the thole pins, the double inner rail is cut to form slots. Crossbeams do not penetrate the hull on this type of *masula*: five thwarts rest on bearers. Normally under each thwart there is a plank tie. Additionally, plank ties may be found in the bow and the stern. These boats are normally equipped with bottom boards – several planks being nailed together over transverse battens to form a removable unit.

The garboards (the lowest side strakes) are sewn onto the posts with wadding, but initially they are not fully secured to the plank-keel: sometimes they are temporarily sewn without wadding, sometimes only lashed into position. Planking-up then proceeds with wadding on all strakes, each plank being cut as required. The shape is determined by holding the plank alongside the one already in position and running a piece of charcoal along the edge. Much trimming takes place with the planks in position. The port and starboard strakes should be as closely matched as possible. The sides are kept apart by the insertion of temporary stretchers.

Once the planking is complete, the plank-keel/garboard seams are permanently fastened. This method is said to allow for adjustments to be made, but, as the garboards are already rigidly fixed to the posts, it is not a convincing explanation.

There is another even smaller *masula* in existence, which was seen nowhere but Bimlipatnam. Less than 5 m in length, it is a paddled craft used for backwater, not sea, fishing. It has the characteristic arrangement of garboard and second strake, however the third strake is not a stealer but a parallel-sided top strake. It has two thwarts, sitting in brackets (as on the *chinna padava*) but with no plank ties under them, although there are ties in the bow and stern.

Sewing method (Figs 5.12, 5.13)

On all these three sizes of boats the method of sewing is identical. A line of holes 10 mm in diameter is bored with a bow drill along the upper edge of a plank. Each hole will be three fingers apart, except on the upper edge of the second-to-top strake and the top strake on the beach seiners, where they are a handspan apart. The line of holes will be one or two fingers' width from the plank edge. On the plank-keel and the posts, the holes are drilled obliquely. The plank to be sewn on will then be matched up. A piece of charcoal is taken and marks are made on this plank so that when it is bored each hole will be directly above a hole on the lower plank.

Once bored, the upper plank is lashed into position at two or three points. These lashings will be untied as the sewing approaches. A length of two-ply

Figure 5.12 Starting a sewing sequence, Bimlipatnam.

coir rope is taken and one end is spliced into the other so that it is double. To the end a short length of nylon fishing line or cotton is tied, and the other end of this is passed through the eye of a large (about 150 mm) metal needle and secured. When fastening strakes to strakes, a start is made roughly amidships (where the upper strake is composed of two planks, it will start at the plank end furthest from the post). Where strakes are being fastened to posts, the sewing commences at the bottom edge and works upwards.

Sewing a seam involves two men: let us call them A and B. One will be inside the boat, the other outside. We need not specify which man is where: although informants invariably insist that the operation commences by passing the needle from inboard to outboard, the reverse – outboard to inboard – also occurs. The man A passes the needle through a hole in the lower strake to B, stopping all the rope running through by one of a number

148

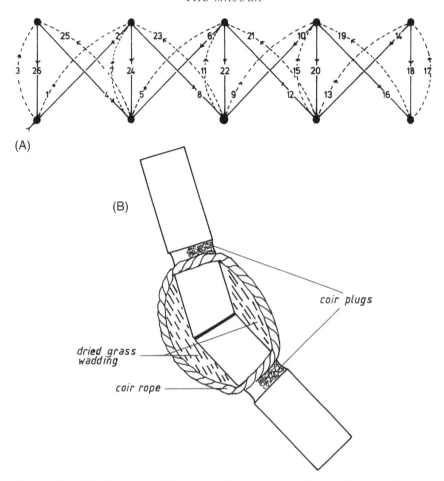

Figure 5.13 Method 1 sewing: (A) the steps; (B) cross-section of the strake-to-strake seam.

of methods. The loop in the rope may be slipped over a metal punch placed in a hole not to be sewn on this run, or a wooden batten may be slipped through it, or a temporary wooden peg may be knocked into the hole to jam the rope. B then feeds the needle through the hole immediately above and, while A hauls down hard on the rope, he knocks a temporary wooden peg into this upper hole. If A has used a peg on the lower hole, he now removes it and feeds the line back through. This stitch is made without wadding and is done to align the planks. Other sewers omit it. The sewing sequence proper now begins and is shown diagrammatically in Fig. 5.13A.

The man B pulls down on the rope, twisting a batten round it. The man A then knocks a peg in. B lays an appropriate amount of dried grass along the seam between the pairs of holes and passes the needle through the upper hole of the adjacent pair. A will pull through and B will knock a peg into

149

this hole. A now lays dried grass on his side and passes the needle back through the lower hole he originally fed through. With B hauling down on the rope, A knocks a peg into the hole. B takes the needle through the hole immediately above, removing the peg from the adjacent hole and, with A hauling down, knocks it into this one. A now takes the rope diagonally to the lower hole of the adjacent pair, crossing the diagonal already made. Passing the needle through to B, A moves his peg from the first hole to this one. This sequence is repeated, extra lengths of coir rope being spliced or tied on and more grass being laid as required, until the post is reached.

The pattern the sewing produces at this stage will be a series of connected crosses (cross-stitch) on A's side and a series of vertical bars connected by single diagonals on B's side. On the top strake seam of the beach-seining *masulas* of both the central and northern sectors (where, as noted, ropes are used instead of grass wadding) the sewing will terminate here, a permanent wooden peg being knocked in to keep the rope tight. On such boats, the series of crosses (A's side) is outboard, which means that the rope must have been originally fed through from outboard.

On all other seams the sewing now 'backtracks'. The rope is taken directly down to the lower hole on A's side. When the needle is fed through to B and A has knocked his peg in, the needle will be passed through the upper hole of the adjacent pair. This sequence is repeated until the starting point is reached. The sewing is then completed by knocking in a permanent peg (which will be chiselled off to make it more or less flush) or a temporary peg if the sewing towards the other post has not yet been done. Holes are plugged with small balls of coconut fibre, knocked into place with a tapering metal punch and hammer. All holes on the smaller *masulas* are plugged, but on the beach seiners this is not done on the top/fifth strake seam or the fifth/fourth strake seam as these are above the waterline.

Only one variation on the pattern described was observed. This was at Visakhapatnam on the top/fifth strake seam of a beach seiner. As on all other beach seiners, this appeared to be 'half-sewn' – that is, the sewing went in one direction only. But whereas, inboard, the pattern the sewing produced was a series of vertical bars connected by single diagonals, outboard the pattern was not a series of crosses but a series of unconnected vertical bars. This was produced by a different sewing sequence, which will be described in the following section.

The Calingapatnam method of fastening the inner rail to the fifth strake on the beach seiners differs from the method used in the northern sector (and the method used in Bimlipatnam, where it is nailed). Pairs of holes are bored through the strake and rail. The pairs are staggered so that one pair bored near the top edge of the rail will be followed by a pair bored near the bottom edge. The distance between pairs is very variable, but not less than 30 cm. The holes in a pair are arranged vertically about 4 cm apart. The rail is fastened by lashings through each of these pairs, which are then tensioned by a single coir rope twisting around each one and forming a zigzag pattern.

Southern-sector masulas

South of Dainaipeta, around the mouths of the rivers Godavari and Krishna, there are no sewn boats. *Masulas* are next found in the vicinity of Kavali (Fig. 5.14) in the villages of Satram, Chennaegipalam, Chakicheria and Ponnapudi (Peddapalam). The boats here are all beach seiners, different in form from the central-sector type, although these villages are in Andhra Pradesh and their users and builders are Telugu (Fig. 5.15).

The boats are around 7 m in length, with length/beam ratios averaging 2.95 and beam/depth ratios of about 1.96. In Ponnapudi it was said that the entire boat was constructed from *Ficus glomerata*. They are comparatively simple craft, with some resemblance to the northern-sector type, although they have six strakes. However, the plank-keel is jointed to the posts in the inverse manner to that described for the northern and central sectors: that is, the lap on the posts is below that on the plank-keel, the heel of the post actually extending below the plank-keel bottom. The posts, which do not taper, are of two parts scarfed and sewn together, with the lap of the lower piece inboard. The garboard strake butts against the posts and the plank-keel

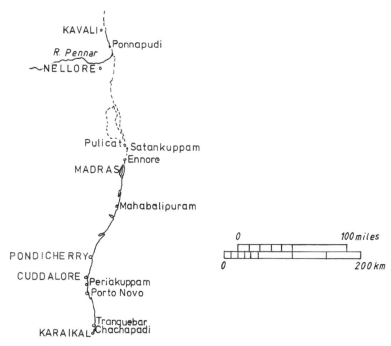

Figure 5.14 Map of southern sector.

151

Figure 5.15 Masula at Ponnapudi.

in the same manner as on the northern-sector boats. The second strake is again a stealer, but for most of its length it has parallel sides. Uniquely among beach seiners all along the coast, there is no stern deck, simply a thwart upon which the steersman sits. He is said to use a paddle instead of an oar, and consequently there is no necessity for the stern-post head to extend above the sheerline. Crossbeams are rectangular in cross-section: some boats have six, others have an additional one just forward of the stern thwart. A substantial inner rail is sewn on – in a pattern similar to that found on the Calingapatnam *masulas*.

However, there is no wadding material outboard on the strake-to-strake seams and the pattern of strake sewing is different from that on the northern- and central-sector *masulas*, although the sewing is the same on the strake-to-post seams. The wadding material used here is coir with a covering of dried grass.

Although in some respects unique, these boats have more features in common with the *masulas* found further south than with those further north. From Madras southwards to Chachapadi all *masulas* are beach seiners, yet they display some degree of variation. Those found between Madras and Cuddalore (Figs 5.16, 5.17) are comparatively large, over 10 m, with an average length/beam ratio of 4.27 and a beam/depth ratio of 2.19. *Aini* (*Artocarpus hirsuta*) is favoured by most builders in this part of the southern sector for the plank-keel, the posts and the lower three strakes, although casuarina may be used for the posts. Some builders prefer *paduk* (*Pterocorpus indicus P. Macrocarpus*) for the plank-keel and lower three strakes. Others use

Figure 5.16 Masula at Madras, returning through surf.

puwarusa (a wood I have been unable to identify) for the plank-keel and posts, *paduk* for the garboards and *aini* for the second and third strakes. Mango is normally used on all boats for the fourth and fifth strakes, although the more costly *red paine* (*Vateia indica*) may be used. The same wood is used for the inner rail. The top strake is of mango or casuarina. Any available wood is used for the crossbeams.

The plank-keel is rectangular in cross-section, about 150 mm broad. On most boats it is the same in depth as the garboards, but on others it is a hand's width. It is fastened to the posts (9 and 7 cubits in length) in the manner described for the Ponnapudi boats. However, although the posts are normally of two parts, they are invariably metal-fastened. Long nails are usual, driven in from outboard with their protruding shanks beaten flat, one pointing up and the other down. One stem post was seen where the two parts were bolted together.

The bottom three strakes are thicker than the others – nine threads – and must be bent by heat. The plank is coated on both sides with castor or *neem* oil along the area in which the bend is to be made and smeared with red sand to prevent splitting. A shallow pit is dug underneath and a fire started. Additional heat is given by a hand-held brand. Once the plank is sufficiently heated, several men will push down hard with a crowbar across the surface of the plank while others haul up the longer end.

One boatbuilder gave an account of a more sophisticated plank-bending procedure. Two holes, 8 feet apart and 3 feet deep, are dug. Into these the ends of two ropes are placed and the holes are filled with stones to jam them in. The plank for bending is placed between the holes, and a log (from a raft)

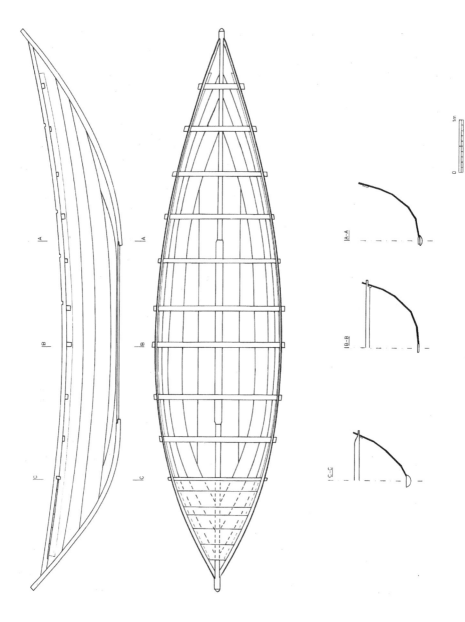

Figure 5.17 Construction drawing of a southern-sector *masula*, measured at Madras.

or a large stone is laid across it and tied down with the two ropes. Some 6 feet away along the plank, a post of about 20 feet is driven 3 feet into the ground. At the top of this there is a pulley and one end of the rope passing through it is tied to the end of the plank. A small pit is dug underneath the plank near the log and a small fire made in it. Again a hand-held brand is used to give additional heat. When the builder judges that the plank is sufficiently warm, he will pull on the end of the pulley rope until the correct amount of bend is achieved. The rope is then tied off and the plank left in this position for two or three hours to cool. The trickiest is the third strake, which must twist as well as bend: a crowbar is used to provide a sideways force.

The pre-bent garboards are positioned so that they butt against the edge of the plank-keel and extend along the stem post 3 cubits and $1\frac{1}{2}$ handspans. The bow on some boats is slightly higher than the stern, on others it is at the same level. On all, the stem post has a greater rake than the stern post.

On the Kavali boats and some Madras craft, the strakes are joined to the posts in the same manner as on the northern-sector *masulas*: they are fastened to the outer edge. This is known as *purakesam* – 'outside' – construction. On other Madras boats the strake ends butt against the inner surface of the post, and this is known as *ulkesam* – 'inside' – construction. We have noted a similar arrangement on the boats of Bimlipatnam, but there this is achieved by placing the edge of the garboard on the top surface of the keel. In the *ulkesam* method, the garboard still butts against the side of the plank-keel but is shaped so that it will 'rise' onto the inner surface of the posts.

The second strake is a tapering stealer. Instead of battens being inserted to keep the sides apart while planking-up proceeds, the correct angle is maintained by a clamping device fixed outboard. Planks on the same strake are joined by sewing over a plain scarf, or more rarely a box scarf.

An inner rail is present on all these boats. This is nailed from inboard so that it is flush with the top edge of the fifth strake. The strake and the rail are recessed to take the crossbeams, which are normally ten, but sometimes nine in number. The crossbeams are rectangular in section and extend outboard. They may be notched, thus forming a double notched joint when in position. The last beam from the bow is frequently shaped. All these boats have a stern deck, four cubits and two handspans wide, which sits on the rail and is secured by the sewing that connects the narrow top strake to the fifth strake.

South of Cuddalore the boats are smaller – between 7.83 and 8.66 m – with an average length/beam ratio of 3.61, rather beamier than those further north. The 'inside' construction method is not known here. There are several important differences between the boats of this area and those of the Madras region. Indian laurel is used for the bottom three strakes; there are seven crossbeams, circular in cross-section; the plank-keel is deeper and the heels of the posts do not extend below it; the fifth strake normally has a hole towards the stern, so that an oar can be inserted to aid lifting when beaching or launching. Most importantly, although they are again six-strake boats, the second and third strakes are both stealers. This is not simply a matter of wood availability: the bottom beam to the maximum beam on one example was 1.78 m to 2.32 m, that is a ratio of 1:1.3. This compares with a ratio of 1:1.67 on the flat-bottomed Ponnapudi boat. Taking the width of the plank-keel, the garboards and second strakes on the more multi-chined northern-sector type and a Madras example, we have ratios of 1:1.9 and 1:1.7 respectively.

Sewing method (Figs 5.18, 5.19)

The unifying feature common to all the *masulas* discussed in this section is the pattern of sewing. The sewing holes are 50 to 60 mm apart, about 30 mm

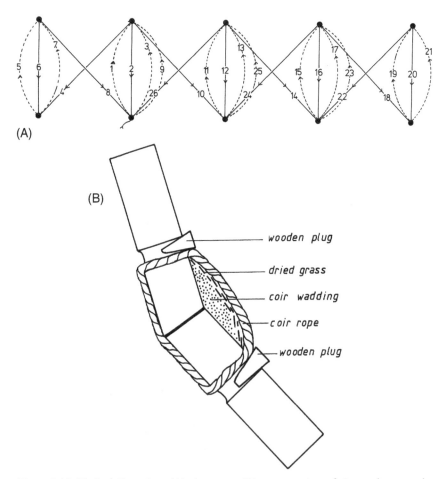

Figure 5.18 Method 2 sewing: (A) the steps; (B) cross-section of the strake-to-strake seam.

Figure 5.19 Sewing near Pondicherry (step 6 in Fig. 5.18A)

from the plank edge and 10 mm in diameter. Dried grass is not used as the primary wadding material. Coir is used throughout, although on many boats it is covered with a layer of marsh grass (*pulu*). The pattern of sewing described above in pages 147–151, which is produced by what we can call Method 1, is also found on these boats, but only where the strake ends are fastened to the posts (where wadding is placed both inboard and outboard). Even here it is common to find a different pattern on the post/garboard seams. However, only one boat (at Periakuppam) was observed which did not display the Method 1 pattern at all: a dilapidated boat said to be in need of rebuilding. Throughout this boat and on the strake-to-strake seams, the plank-to-plank seams and usually on the post/garboard seams of all boats in the southern sector, a different sewing method is used, which we can call Method 2.

On the inside of the hull, the appearance of the sewing is the same as on boats of the northern and central sectors – a cross-stitch with vertical bars – but outboard the pattern produced is of unconnected vertical bars. Where Method 2 is used, the outboard seams are not wadded with coir, grass or any other material, except on the post/garboard seam. However, if the boat is being resewn and the planking is particularly badly worn, then the midribs of palm leaves, bicycle-tyre inner tubes or even old flip-flop sandals may be placed over the seam outboard at such points. These do not form a continuous bow-to-stern wadding.

For sewing, a doubled length of coir rope is again used, which is fastened to a metal needle in the same manner as previously described. Temporary pegs are again used to keep the line taut as the sewing proceeds. The sequence is shown diagrammatically in Fig. 5.18A. Let us call on A and B once more to perform the operation. In Method 2, the sewing always begins by passing the needle through a lower hole inboard, where we will position our man A. Man B takes the line outboard and feeds it through the hole immediately above. A lays a wad of coir over the seam and passes the needle back through the starting hole. B takes the line back up to the hole above and feeds through. A takes the line diagonally to the lower hole of an adjacent pair in the opposite direction in which the main sewing run is to proceed. B takes it up through the hole immediately above and A takes it straight down again. B then passes the needle through the upper hole again and A takes the line diagonally across to the starting hole. B takes the line up to the hole immediately above and the sewing is complete on these two pairs of holes. The sewing now continues by A taking the line diagonally across to the lower hole of the adjacent pair in the other direction. B takes it up through the hole immediately above, A takes it down and B takes it up again. This sequence is then repeated until the stem or stern post is reached. At this stage, excluding the first two pairs of holes, the pattern produced will be a series of connected 'N's – vertical bars with single diagonals – inboard and unconnected vertical bars outboard. The line is then taken back along the run, A taking it from the upper hole of one pair to the lower hole of an adjacent pair and B taking it from the lower to the upper hole of a pair, until the starting hole is reached.

The seams of the third and fourth strakes are not normally plugged, and the seams of the fourth and fifth and the fifth and the top strake are never plugged; nor are the post/strake seams above the waterline. The holes are plugged, not with balls of coir but with tapered wooden pegs.

As noted, Method 2 is normally found on the post/garboard seam but Method 1 on all the other post/strake seams. However, a few boats display Method 1 on the posts from the top strake to the plank-keel. The reason for this is not clear. Method 1 may be used on the posts because these must bear the brunt of the breakers and handling when the boat is being beached or launched, and therefore extra binding and wadding are needed. But why is it not normally continued onto the garboard, even though wadding is used here? Boatmen who have Method 1 right to the plank-keel on their boats say that it is simply for extra strength and that other boats would have it if their owners could afford it. This is not very convincing, but no alternative explanation can be offered. The difference does not appear to be related to the *ulkesam* and *purakesam* methods of construction.

As on the northern and central-sector beach seiners, the top strake sewing holes are a handspan apart and the seam is 'half-sewn': the sewing does not backtrack, but here it is an abbreviated form of Method 2, producing vertical bars outboard and the connected 'N's inboard. No wadding material is used on this seam.

Propulsion and steering, beaching and launching

With the exception of the three-strake *masula* of Bimlipatnam, which is paddled, and the ferries of Navalarevu, which are poled, all *masulas* are rowed, although the four-strake boat of the central sector is also sailed.

In Tamil Nadu, each boat has six or eight rowers, each man having an oar, and one steersman. In the area between Madras and Cuddalore, the crossbeams are used as thwarts and stretchers. The bow oarsman sits on the first crossbeam, rowing to starboard, with his feet on the second crossbeam. One oarsman sits on the second crossbeam, rowing to port. Two men sit on the third beam, two on the fourth and commonly two on the fifth. South of Cuddalore the rowing arrangement is the same, but the crossbeams, which are fewer in number than on the Tamil Nadu boats further north, are used only as stretchers. Removable thwarts are tied to the top strake, with the foremost hard up against the stem post. On all these boats, the thole pins are similar and are secured by drilling one or two pairs of holes in the fifth strake and lashing, fitting into recesses cut in the top strake, just forward of the crossbeam used as a stretcher by the oarsman. The oars lie on the forward side of the pins and are attached by a loop of coir rope twisted around the shaft of the oar and slipped over the top of the pin: that is, the oars work against this rope, not against the pin.

The oars are approximately 3.25 m in length, with a blade length of around 0.40 m. The blade is separate and has a 'handle' to which the shaft is lashed. In addition two pairs of holes are bored through the blade and further lashings to the shaft are made. There is quite some variation in blade shape, which does not seem to be attributable to geographical areas. Even on a single boat the blade shape may not be standardised. The steering oar is larger – about 4.75 m – with a blade length of around 0.80 m. It may or may not have a 'handle'. It is secured to the port side of the stern-post head in the same way as the oars are secured to the pins, although a wad of old net is often tied to the shaft or to the post to prevent it slipping.

As noted in the preceding section, the *masulas* of Ponnapudi do not have a steering oar but a hand-held paddle. The blade shape of the rowing oars is the common 'ace of spades' form. However, there are six oarsmen and four oars. The bow oarsman perches on a thwart hard up against the stem post, pulling to starboard. Two men work the second oar to port, two men work the third to starboard and one man pulls the fourth to port. The oars are of different lengths. On one example, the bow oar measured 2.81 m, the second oar 5.10 m, the third 5.75 m and the fourth 4.20 m.

A somewhat similar arrangement to this is found on the boats of the northern and central sectors, although they may have up to thirteen oarsmen.

One or two men work the bow oar to starboard, up to three men work the second to port and up to four men are found on both the third and fourth oars. However, in both the northern and central sectors there is a small number of boats with two oars on the third station, each pulled by one man, and no fourth station. A boat at Calingapatnam was also found where the bow oar worked to port.

As in Ponnapudi, the length of the oar in Visakhapatnam varies according to its position, but in both the northern and central sectors the blades of the oars are asymmetrical along a longitudinal axis, the precise shape differing from oar to oar.

Throughout the coast, the side of the oar blade which has the stock on top is the side that pushes against the water, even though there would seem to be an advantage in having the stock behind the blade as it pushes. Oars of the northern and central sectors are secured to the thole pins in the same manner as in the southern sector, but the pins are somewhat more substantial.

The four-strake boat of the central sector is equipped with three oars, similar in shape to those of the beach seiner. However the thole pins are not lashed into place but fit into sockets.

Oars are not just for rowing. In the northern and central sectors they are tied to the crossbeams of the beach seiners when the boats are being taken to and from the water, the overhanging ends being used as yokes. The boat is taken, half lifted, half dragged, in a straight line to and from the shore. As noted, south of Cuddalore in the southern sector, boats have a circular hole cut in the fifth strake towards the stern so that a lifting oar can be slipped through. However, around Madras the projecting beam ends are used as yokes. Here the boat is not taken by a straight lift but pivoted on the heels of the posts and 'walked' up or down the beach.

Lacking crossbeams, such methods of moving the boat on land cannot be used for the four-strake *masula*. Instead, at the bow and stern a rope loop is found on each side through which an oar can be inserted.

The sail of this boat (Figs 5.20 and 5.21) is a boomed trapezial lug (Doran, 1981: 40), identical to that used on some *teppas*, and it is rigged in almost the same manner. A beam, circular in cross-section, is laid across the sheer between the first and second thwarts and tied down by lashing through pairs of pre-drilled holes in the top strake. The mast, about 2 m in length, is slotted into a loose mast step jammed into the bottom of the boat and rakes forward. A loop of rope is hooked over the end of the beam and over the crosstree of the mast on each side. On some boats the backstay runs from the crosstree; on others it runs from the masthead back to a hole in the top strake just forward of the last thwart. Some have a halyard, running through a hole in the masthead and tied off just aft of the beam. Others simply have the yard lashed to the mast. The foresheet of the yard is made fast to the foremost thwart and the two other sheets, from the yard and boom ends, are tied to holes in the top strake, within easy reach of the aftermost thwart. The

Figure 5.20 Chinna padava at Paradeep.

forward end of the boom is fastened to the beam. A sail noted in Visakhapatnam measured $12.44m^2$ (Mohapatra [1983: 17] notes a triangular sail and lists dimensions that would give a sail area of $13.52m^2$).

Decoration

Carved stem-post heads are rare. Only two examples were observed, both similar: one at Elliot's Beach, Madras, the other near Tranquebar. No boat seen had the coconut shell or pot decoration shown in Hornell's *Water Transport* photograph (1946: frontispiece), although the figure painted on it is similar to that found on the strakes of all the Ponnapudi boats (see Fig. 5.15). This motif was said to ward off the evil eye.

Again near Tranquebar several boats had green pennants on short staffs tied to the stem-post head (indicating that the owners were Muslim). This

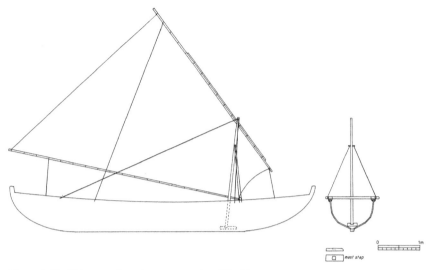

Figure 5.21 Chinna padava rig, measured at Visakhapatnam.

practice was not observed elsewhere. In the northern sector, however, it is common to find a small straw garland hanging from the post: on one boat this was replaced by a wooden doll.

The boats of Tamil Nadu have their third and fourth strakes coated with a mixture of pitch and tar (in the ratio of 2:1). This is said to be a preservative measure and is applied after construction with a 'brush' – a handful of coir – in such a way that it does not cover the seams. The garboard and second strakes are not coated. This mixture is not used elsewhere, although black paint is sometimes found, for example in Visakhapatnam. Completely undecorated boats are to be seen in the northern and central sectors, but normally at least the second-to-top strake is painted. There are no obvious rules about colour. The common practice between Madras and Cuddalore is to divide the fifth strake into three longitudinal bands, often with the upper and lower of one colour and the central band white, with single colour areas at the strake ends. Elsewhere, strakes are normally of a single colour. Throughout the coast the top strake is more often left plain than painted. In the central sector, some beach seiners have slanting lines of paint (often of alternating colours) on the top and fifth strakes, but none were seen that exactly conform with Hornell's photograph (1946: frontispiece).

The painting of numbers on boats, which the drawings of Hunt and Folkard indicate, has almost ceased, presumably because of the ending of the ship-to-shore function of the boat. Only in Calingapatnam and Paradeep does the practice persist, seemingly to comply with local regulations. Boats are not given names. At Edumanipelam I came across a boat on which was written in Telugu *'jalow usha'* – 'water dawn' – although I was assured that this was not a name for the boat. Owners may paint their own names on their boats, but the initials in Roman script 'MGR', which at least in the early 1980s a few Tamil Nadu boats displayed, belonged to the state's Chief Minister, M.G. Ramachandran.

Similarly, motifs painted on boats may be political symbols. Many Tamil Nadu boats had a flower painted on them – the symbol of MGR's AIADMK party. However other motifs may be purely decorative. A pendant motif is common in the northern sector: I do not know if it is a political symbol. Only once did I find a boat with a boat painted on it. This was in Puri.

The oculus is not as rare as it may have been in Hornell's time (1920: 174): perhaps it has undergone a revival. It is very common in the northern sector, painted on either side of the bow on the fourth strake (although on one boat in Sonapur eyes were painted on the third strake; this was also unusual because they had eyelashes). The oculus is not found in the central sector but is quite common from Elliot's Beach, Madras (though rare on the Marina Beach), southwards to Mahabalipuram. It may be chiselled on, giving it a diamond shape, but is more commonly painted on the fifth strake. The painted version is given an eyebrow and is often accompanied by a curved white line painted on the black fourth strake. Hornell (1923: 320) could find no explanation for this line, although he records a similar design on some of the canal boats of the Krishna-Godavari complex. I can do little better. It is called the *messai* – 'moustache' – and I was told that it signifies that the boat is male. However, asking other fishermen about the gender of the boat, without reference to the *messai*, I was told that it was neuter. In the northern sector, incidentally, the hanging garlands on the *masulas* were explained as *nothus* – the nose decorations of women – indicating that the boat is female! It is likely that the *messai* is a traditional motif, the original significance of which has been forgotten. It is worth noting that somewhat similar oculi have been found on representations of boats of Ancient Egypt (Hornell, 1923: plate XIX). One interesting variation seen on the Marina Beach, Madras, was an oculus transformed into a flower.

Conclusions

Having examined the range of *masulas*, it is clear that it is impossible to advance a more detailed definition than that presented in section 5.5 if a definition is required to cover all *masulas*. Beyond being frameless, fine-sterned sewn plank boats, the only attribute they all share is Method 1 sewing somewhere on their hulls.

There is little further scope to allow a more extensive 'polythetic' definition (Clarke, 1978): that is, to treat *masulas* as a class of boats sharing a large number of attributes but where the presence of no one attribute is an essential condition of class membership. A polythetic definition of the type could run: a frameless, narrow and deep (L/B = >3.75, B/D = <2, McKee 1983: 79), six-strake sewn boat between 8 and 10 m in length, which is rowed and used for beach seine fishing.

Masulas can be subdivided according to function, which would give us beach seiners, gill netters, backwaters fishing boats and ferries. Numerically, the first category is enormous, the others almost insignificant. The boats could equally well be classed according to the method of propulsion: rowing (which could be further subdivided into those boats with as many oars as rowers and those with fewer), sailing and rowing, paddling and poling. These two systems do not entirely coincide: the ferries of Navalarevu are poled; the Gopalpur ferry is rowed. However, this account has been primarily concerned with the construction and form of the *masula*, and neither of these systems sheds much light on structural differences. In fact they hinder: the ferries of Navalarevu are structurally identical to the beach seiners of that village except they lack a strake above the crossbeams, and, as can be seen from Table 5.4, the Visakhapatnam seiner has more attributes in common with the gill netter and backwaters fishing boat than with, for example, the Ponnapudi beach seiner.

The interpretation of the *masula* complex that has been presented here is that the boats fall into three basic constructional groups. These can be differentiated by combinations of several attributes but can be most easily differentiated by a combination of just two attributes: the form of the second strake and the method of strake-to-strake sewing. On the boats of the northern sector and the southern sector the second strake is a stealer. On the boats of the central sector it is a tapering, but full, strake.

This divides *masulas* into two groups. But if the pattern of strake-to-strake sewing is examined, a different division is found. On the boats of both the northern and central sectors, wadding is found both inboard and outboard, held in place by the double web that Method 1 produces. On the boats of the southern sector, wadding is only found inboard, held by the single web that Method 2 produces. The fourth permutation – where the second strake is not a stealer and Method 1 sewing is not used on strake-to-strake seams – does not seem to exist.

The similar way in which the plank-keel and the stem/stern posts are joined and the similar rowing arrangement are further attributes that distinguish the northern- and central-sector boats from those of the southern sector.

However, there is a broader context into which the *masula* fits. In *A Handbook of Sewn Boats*, Prins (1985), acknowledging the dearth of published information on the *masula*, tentatively consigned it to his class P9. It shares this class with the other sewn boats of the Indian Ocean, such as the *mtepe* of Kenya, the *beden* of Somalia and the *oruwa* and *paruwa* of Sri Lanka. However, the attributes required for membership of this class are flush-laid planking, continuous sewing (as opposed to lashings), plugs stopping the sewing holes, and treenail (or dowels) between the planks. The *masula* does not have the last attribute – treenail between planks – and therefore it belongs in Prins's class P10 (as does the *paruwa* – see Chapter 6). This class contains all the craft with flush-laid planking, continuous sewing and plugs but without treenail. According to Prins's information this amounts to the Samoyed boat of Western Siberia and certain Polynesian craft. This does not lead us very far: even a cursory examination reveals major constructional differences between these craft and the *masula*. The Samoyed boat is framed and is without wadding; the Polynesian craft use solid battens behind the sewing; the *masula*, as we have seen, is frameless and uses a non-rigid wadding behind the sewing, as it appears do all the other Indian Ocean sewn vessels.

A Handbook of Sewn Boats reveals the enormous variety of patterns of sewing patterns around the world, yet, from Mombasa to Madras, on the *mtepe* of Kenya, the *beden* of Somalia (Chittick, 1980), the *sanbuq* of Oman (Facey, 1979), the dhow *Sohar* built by Laccadive islanders for Tim Severin (Severin, 1982), the boats of Kerala, and the *oruwa* and *paruwa* of Sri Lanka (see Chapter 6), the sewing pattern appears to be identical. Those sewn boats from the Indian Ocean recorded by Pâris (1843) – the Muscat *béden*, the Sri Lankan *yathra dhoni* and the boats of Mangalore (as well as his drawings of the *masula*) – again all show this one pattern. This is a web on one side of the hull and vertical unconnected bars on the other. It is the pattern described in this chapter as Method 2.

The western limit of this 'Indian Ocean pattern' is the coast of East Africa. It does not travel inland – the canoes of Lake Victoria being lashed together, not truly sewn. The eastern limit seems be the Chittagong *balam* (Greenhill, 1971: 113). However, this uses split bamboo for both wadding material and sewing material, whereas coir rope seems to be the preferred material for sewing in all other areas (unless superseded by artificial fibre ropes).

Yet within this Indian Ocean complex, only the *masulas* of the southern sector of India's east coast (as defined here) are true members. The predominant sewing method in the central and northern areas is a double-web method for which there is no evidence anywhere else in the Indian Ocean.

There is no evidence that the double web has ever been more widely spread on the east coast of India than it is today, although Pâris's illustrations suggest that the single web has moved from outboard to inboard on the side planking of southern-sector boats. It is therefore likely, but not certain, that Method 1 has evolved from Method 2. Perhaps the significant difference between the northern- and central-sector boats that use Method 1 and the southern-sector boats that use Method 2 is the wadding material: expensive coir wadding needs to be used on only one side of a seam and therefore only one web is required; cheap grass is not as effective and needs to be used on both sides of the seam, therefore requiring the double web.

Acknowledgements

The research on which this chapter is based was funded by the National Maritime Museum, Greenwich. Throughout the fieldwork periods I was accompanied by an interpreter, and I am greatly indebted to T.K. Chalapathy, Kanagaraj David, M.S. Jagannathan and B. Panigrahi, who each fulfilled this role and acted as invaluable companions.

6

THE *MADEL PARUWA* OF SRI LANKA – A SEWN BOAT WITH CHINE STRAKES

Eric Kentley

On completing the research on the *masula* documented in Chapter 5, it was a logical next step to study another sewn boat in the same approximate geographical locality to discover what similarities, if any, might exist. The *madel paruwa* of Sri Lanka was an obvious choice. Not only is it a sewn plank boat, it is also, like the *masula*, a beach seining boat. The research on the *madel paruwa* was undertaken in January 1986, when the coastline between Kalpitiya and Kirinda (Fig. 6.1) was surveyed specifically to locate and record variations on the *madel paruwa*. Aspects of the results were originally published in Kentley and Guneratne (1987).

Features of the *madel paruwa*

On most flat-bottomed plank boats, for example the dory, the sides of the boat are joined to the bottom by fastening the lower part of the inboard face of the lowest side plank to the outboard edge of the outermost bottom plank. However, there are some flat-bottomed craft where a shaped longitudinal timber – shaped in cross-section from a 'C' to an 'L' shape – is located between the bottom strakes and the side strakes. This timber has been termed a 'bilge strake', 'transition strake', 'chine girder' and 'ile' by archaeologists and ethnographers, as well as 'chine strake'. In this chapter I have adopted the term 'chine strake' in preference to the others as it locates the feature and implies that it runs longitudinally.

The *madel paruwa* (plural *paru*) of Sri Lanka is a boat with such a feature (Figs 6.2, 6.3). The word *paruwa* translates as 'floating aid' and is applied on the island to three functionally distinct craft: a river-going passenger ferry, a river-going sand collector and transporter, and a surf boat used for beach seine fishing. The latter is distinguished from the other two by the prefix *madel*, literally 'big net'. The beach-seining operation on the east coast of India has been described in detail in Chapter 5 and is essentially the same on the west coast of Sri Lanka.

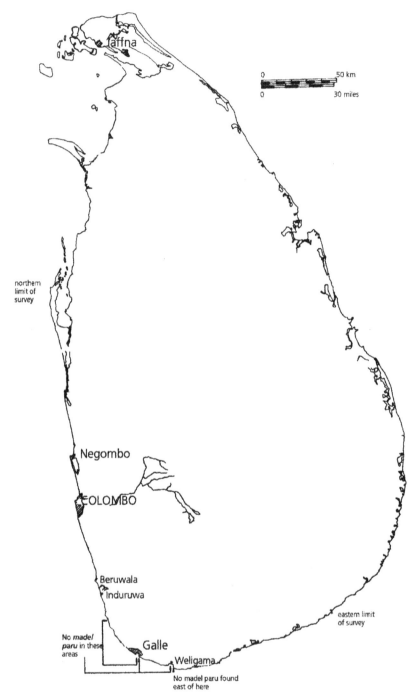

Figure 6.1 Map of Sri Lanka showing places mentioned in the text.

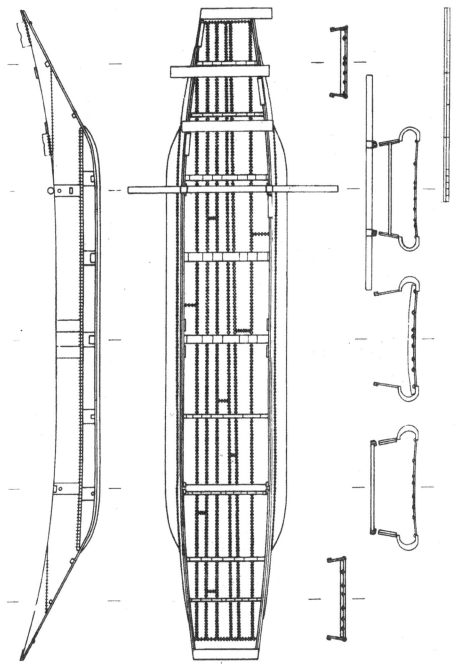

Figure 6.2 Construction drawing of a *madel paruwa*, measured at Beruwala.

Figure 6.3 Madel paruwa at Pandara, near Weligama.

As with probably all traditional craft not built with the aid of formal plans or according to rigid formulae, the *madel paruwa* is not a standard product. The survey between Kalpitiya and Weligama revealed considerable variation. Nevertheless, all *madel paru* have sufficient attributes in common to distinguish them from other boat types and to arrive at a general definition:

- all have chine strakes
- all are fastened principally by sewing
- all are scow-ended, like a punt
- most are concave-bottomed in cross-section, rather than truly flat-bottomed (they have what could be termed 'reverse deadrise')
- most have a transverse bar fitted for carrying the boat
- most have transverse boards fitted at the bow and stern – end boards – to which the ends of the bottom strakes are fastened
- all are propelled only by rowing

De Zeyla (1958: 5) published a curious photograph of a *madel paruwa* which at first glance appears to have a mast and sail and which he captions 'Beach seining *paru* [*sic*] at Negumbo fitted with outrigger and sail. Many *parus* are much larger and not fitted with outrigger or sail'. This would be an exception to the last attribute listed above. However, my interpretation of this photograph is that there is an *oruwa* concealed behind the *paruwa* and it is to the former that the sail belongs. I found no physical or verbal evidence of a sailing *paruwa* or, incidentally, of a *paruwa* with an outrigger having been used in Negombo.

Construction

The boats are normally around 10 metres in length, built without frames but with floor timbers. *Jak* (*Artocarpus intergrifolia*) is the preferred wood for both the chine strakes and planking, but mango (*Mangifera indica*), which is cheaper, may be substituted. Construction begins by shaping the outside of a single log from which the chine strakes will be created. In Weligama and Galle the chine strakes are exceptionally broad – wide enough for the steersman to stand on – and, unless a very broad log is readily available, so that two can be fashioned from it, it is normal for each chine strake to be carved from a single tree trunk. Elsewhere the log provides both chine strakes. The log is shaped to an almost circular cross-section, but with the top and bottom flattened. The top edge is not normally horizontal in profile: the centre should be some 3 inches (76 mm) lower than the ends. There seems to be no formula for the longitudinal curve at the ends: this is said to depend on the waters in which the boat will operate. However, the curve at the after-end should be slightly steeper than that at the fore-end (Fig. 6.4).

In the case when one log is to provide both chine strakes once the shape is achieved, the log is smoothed with a plane and then sawn down the centre longitudinally. The next step is to hollow out the two half-logs, which is done with axe and adze. Each half is hollowed so that it is thicker at the bottom than at the top. In order to gauge thickness, a few holes are bored through so that a thin stick can be inserted and the depth measured.

On a larger number of *paru*, the inside of the chine strake is carved so that it is particularly thick at the points where floor timbers will meet it. This is said to provide additional strength, but it seemed from observation that this feature was being dropped on new craft in favour of specially carved strengthening pieces which conform exactly to the inboard shape of the chine strakes and are nailed into position from outboard. These are sometimes notched to take the ends of the floor timbers and prevent them riding up.

The planking for the bottom strakes and the single side strake – the washstrake – is nine threads thick, that is $1\frac{1}{8}$ inches (29 mm), and is obtained through-and-through cut (i.e. cut tangentially to the growth rings) and waney-edged (i.e. including sapwood at the edges) from the local saw mills. To give himself a straight line to saw along, the boatbuilder uses a type of chalk line, but using a solution of carbon from old batteries in water instead of chalk. The number of bottom strakes varies according to the breadth of the available planking. However, one does not normally find a single plank running the full length of the boat, but instead several planks within each strake. The normal practice is, when they are in position, simply

Figure 6.4 Completed chine strakes, Beruwala.

to butt-joint and secure by sewing. The builder ensures that such joints are staggered.

There is one more operation that must be performed before the assembly of the boat can begin. The planks are too thick to be forced to conform to the curves of the chine strakes: they must be bent by heat. Some boatbuilders bend all the planks in two operations (one for the bow planks and one for the stern); others bend each plank individually. The basic technique is nevertheless standard. Two upright posts are driven into the ground and a metal bar is placed and secured across the top. The area on the plank to be bent is coated with coconut oil (in some areas clay is also applied to prevent splitting). The plank (or planks) is then propped up on the metal bar. The lower end is weighted, the upper end not necessarily so. A fire will then be lit underneath and slowly the plank will begin to bend. A template, which may be a metal rod or the midrib of a coconut leaf, seems commonly to be used to check the shape against that at the end of the chine strake.

The main components are now ready and a building platform is constructed. This consists of three or four stocks – two upright poles with a crossbar – about $2\frac{1}{2}$ feet (0.75 m) above the ground, with stocks of twice this height at each end. One practice is to place the two chine strakes on the stocks and start to sew on the bottom strakes from each side, working towards the centre (Fig. 6.5). The alternative is to sew the bottom planks together first and then attach the chine strakes.

Sewing

The sewing is done through pre-drilled holes. There is considerable variation in the diameter of these holes and their distance apart, but they are always bored so that each aligns squarely with a hole on the adjacent plank or chine strake.

The sewing operation begins by laying *dumala* – a resin obtained from decayed trees which is heated in coconut oil to make it pliable – on the inboard surface of the two strakes, which are flush-laid. On top of this, wadding material is placed. South of Colombo, coconut fibre (coir) covered with dried leaves from the coconut tree is used. To the north, they no longer use coir but the leaves alone. Allegedly, bicycle-tyre inner tubes can also be used here, but this was not observed. The wadding material, inboard only, is

Figure 6.5 The raised chine strakes.

held in place by the sewing. The 'needle' used is a piece of bent wire, or even a piece of straw, to which a small loop of string is attached. And it is to this loop that the sewing line is attached. In Weligama and Galle coir rope is still used. Elsewhere it has been superseded by nylon net string. The line is usually doubled, but in Beruwala it is trebled.

The technique of actual sewing is completely standard. It is a two-man operation: one inside and one outside the hull, which means most of the time one man is above the other. The sewing of two strakes usually begins amidships and works towards one of the ends. The line is kept taut by knocking a temporary peg into the hole the line has been passed through while the man on the other side of the hull hauls. The man inboard lays the wadding material as he goes, beating it flat with a hammer. The sewing produces a 'criss-cross with verticals' web inboard and disconnected bars outboard. This is identical to the pattern found in India on the southern-sector *masulas* discussed in Chapter 5 and the steps taken to achieve it are identical to those shown in Figure 5.18A. In the early nineteenth century, F.E. Pâris (1843: 24) recorded that grooves were cut on the outboard faces of the planks to recess and thereby protect the stitching – a common characteristic among Indian Ocean sewn boats – but this practice seems to have disappeared completely (nor is it found on the *masula*).

Sewing of the bottom planks is completed by plugging the holes with wooden pegs or small balls of coir. The floor timbers are now fitted. The *paruwa* is not quite a flat-bottomed boat, but unlike all other nearly flat-bottomed boats (with the exception of some modern power boats, and some Severn trows which Finch [1976: 121] noted 'were actually constructed with floors which sloped slightly upwards towards the keel . . .') it is not convex in cross-section but concave. This feature gives some protection to the sewing on the bottom strakes when the boat is being beached or moved and must also inhibit the movement of free water across the bottom, which could have a destabilising effect. The degree of concavity is variable but is achieved by inserting shaped floor timbers across the bottom. To get the planks to take up this shape, a prop or car jack is forced underneath and the timbers are lashed down tightly with a running stitch through holes bored in the bottom strakes. These floors usually have limber holes cut in them, and on the bottom strakes there is usually one or more drainage holes stopped with a wooden peg or bundle of coir.

Fitting end boards and crossbeams

At this stage the boat is normally removed from the stocks, as the remaining work is more conveniently done with the boat on the ground. The transverse end boards are attached. Although these have wadding inboard in the same manner as the strakes, the seam is reinforced outboard with a wooden batten, held by the 'vertical' bars of stitching. The floor timbers between the ends of the boat and the chine strakes are now fitted, but these are flat timbers, not concave. On some *paru* around Negombo there is a transverse board in the place of the aftermost floor.

The washstrake is now sewn on, in the same manner as the bottom strakes but commencing outboard, so that the wadding and criss-cross pattern are also outboard. The washstrake is usually a single plank with a triangular piece at each end but can also be made up of a number of planks. There is considerable variation in the degree of tumblehome (a term meaning that the sides of the boat curve inwards – the opposite of flare: see Glossary). On some boats a narrow rail is nailed onto the outer top edge of the washstrake; on others it is found on the inner edge; on others on both edges; and on yet others there are no rails at all. Similarly, some *paru* have one or more crossbeams keeping the sides apart, others do not. The most elaborate crossbeams are found on the *paru* of Balapitiya. Each boat has three, each of which curves up towards the centreline.

A further characteristic of the *madel paruwa* is the transverse bar towards the bow which is used to carry the boat. This may be attached in one of three ways: it may be laid on the washstrake and lashed through holes and to a crossbeam immediately below; it may be slotted through holes in the washstrake and lashed down; or special sockets may be sewn on top of the washstrake and the bar slotted through these. The latter is the practice north of Colombo, and the bar is not fixed but can be removed to make loading and unloading the net easier. It is not, however, removed when going to sea.

This bar alone is not sufficient to lift the boat, and one or more pairs of holes are bored through the washstrake towards the stern through which an oar can be inserted. Where these holes are bored, strengthening pieces are nailed on, running the full depth of the washstrake inboard or outboard or both.

Propulsion

The method of propulsion is by rowing, although in Weligama and Galle poles are also used in the shallows. Blocks of wood are sewn onto the top edge of the washstrake, reinforced with battens, and through these rope grommets are fastened. The method of rowing also has some variations. In some places there is one man per oar; elsewhere there is more than one man for each oar. Commonly there are four oars, but sometimes five. The oarsmen sit on benches (thwarts) laid across the washstrakes, using the next bench as a stretcher, but additional rowers may stand.

Again, there is some variation in the method of steering. In Beruwala a hand-held paddle is used, but the most common method seems to be for the steering oar to be pivoted on a rope grommet on the starboard quarter. North of Negombo this seems to be giving way to the steering oar pivoting on a centrally mounted block.

The outrigger *paru*

The most striking variation of the *madel paruwa* is seen at Induruwa (Fig. 6.6). Here the *paru* have a single outrigger float, identical in form and method of attachment to that found on the *oruwa* (plural *oru*), the logboat (usually extended) which is the most common traditional craft of the Sri Lankan coast (Fig. 6.7; Kapitan, 1991). The outrigger is found on the *paru* where a special technique of beach seine fishing is found and the *paru* are significantly narrower. The *paru* at Induruwa work in pairs, each taking half the net. The two meet at sea and join together with poles. This enables them to lace up the two halves of the net, the main body of which can then be shot. To make this possible, one of the boats has the outrigger on the port side, the other on the starboard side. Unlike the *oruwa*, the *paruwa* is not double-ended in the strict sense of the term.

Figure 6.6 Outrigger *paruwa* at Induruwa, with roof to cover the net when not in use.

Paru with outriggers are also found between Ratmalana and Wadduwa (and possibly a little further south). At Moratuwa, just south of Ratmalana, two *paru* in use there both had their outrigger floats on the port side. The explanation was that, although the two-boat technique is used here, they also shoot nets independently. If an especially abundant catch is anticipated, a particularly large net will be brought to the beach and the outrigger on one of the boats will be switched to the other side.

The *paruwa* in the Indian Ocean context

One of the many characteristics the *paruwa* has in common with the *oruwa* is the method and pattern of sewing. As discussed in Chapter 5, they both fall

Figure 6.7 Small *oruwa*, Kalpitiya.

within the Indian Ocean sewn boat group, of which the Indian *masula* is a fringe member. What the *masula* and the *paruwa* also have in common is the absence of a feature that seems to be standard on all other Indian Ocean sewn boats (Prins, 1985): neither employs treenails as an additional device for locating and/or securing the strakes. Given that these two craft perform the same function, the geographical proximity of Tamil Nadu to Sri Lanka and the history of Tamil migration to the island, one might perhaps expect a greater degree of similarity between the two craft. However, the differences seem greater than the similarities. Although the *masula* is a relatively flat-bottomed boat, it is fine-ended with stem and stern post. It has no chine strakes: the outer bottom and the lowest side strakes are directly fastened. *Paru* are used principally by Sinhalese fishermen, although some Sri Lankan Tamil fishermen also use them.

177

The *masula* is a unique craft. There is nothing on the east coast of India which even remotely resembles it. In Tamil Nadu the predominant fishing craft is the *kattumaram* – the lashed log raft, which incidentally is also used by Tamil fishermen in Sri Lanka. Logboats exist but are not manufactured in Tamil Nadu: they are imported from the state of Kerala on the west coast. Sri Lanka, in addition to the *kattumaram*, has another type of log raft, which is pegged together: the *teppam*. The Tamils of the island also have a fine-ended outrigger-less logboat, the *vallam* (quite different from the *vallam* described below in Chapter 7), which can be used for beach seining. But the predominant fishing craft among the Sinhalese is the *oruwa*.

The *oru* are a distinctive group of logboats, ranging from 2 to 10 or more metres in length. The log has a marked tumblehome, and its ends taper slightly and sweep upwards but on most types are left quite blunt: they do not form a point and are scow-ended. Hornell (1943: 46) reasoned that the *paruwa* was derived from the *oruwa* because of the presence of chine strakes: he suggested that the *paruwa* was simply an *oruwa* cut down the middle with planking inserted between the two halves. Indeed, Pâris (1843: 24) states that old *oru* were used to make the sides of *paru*, although this no longer seems to be the case. Given not only the presence of chine strakes but also that the chine strakes have a similar tumblehome to that on the *oruwa*'s log underbody, and that both craft are scow-ended with their ends closed by raked end boards, it does seem possible that the *paruwa* is derived from the *oruwa*. It appears to be a logboat builder's – specifically an *oruwa* builder's – solution to a requirement for a boat with extra capacity. The shape of the log underbody does not allow for much vertical extension, therefore transverse extension was developed. There is a counter argument that the chine strake is derived from the hewing out of side-strake planking rather than from log hollowing, but Sri Lanka does not appear to have a strong tradition of building plank boats.

The *madel paruwa* is not the only boat with a chine strake on the island. Sand-carrying *paru* (Fig. 6.8) on the River Kelani near Colombo, although they are not sewn but nailed and bolted and are heavily framed with no sheer, nevertheless have a similar general plan to the *madel paru*. They are scow-ended and taper slightly at the bow and stern. Although they are quanted/poled and not rowed, their steering arrangement is identical to that found on the *madel paru* north of Colombo. Most importantly, they almost all have chine strakes. However, because of the presence of ceilings and bottom boards it was not possible to discover how they were fastened. The only such *paruwa* which could be found under construction at that time was one of only two observed that did not have chine strakes! Instead the builder was intending to fasten a substantial external batten, roughly rectangular in cross-section, to the bottom edge of the lowest side strake. This was observed on a boat in use and suggests that the chine strake is regarded primarily as a strength member, not simply an intermediary member between sides and bottom.

Figure 6.8 Sand-carrying *paru*, near Colombo.

No passenger *paru* could be located during this survey, but Hornell's description (1943: 47–8) implies that it also has, or had, chine strakes and may be sewn.

Tamil Nadu has perhaps 'failed' to develop a craft similar to the *paruwa* because it lacks a logboat building industry. This is true of most of the Indian Ocean. Arabs traditionally import logboats from the Malabar coast (Facey, 1979: 151). In those areas where logboats are made other than Sri Lanka, such as Tanzania as well as Kerala, they are sharp-ended with minimal washstrakes and without the tumblehome of the *oruwa* log base. Similarly, although outriggers are known, for example in Tanzania, they are double and are connected to the hull in quite a different fashion from the Sri Lankan practice (Hornell, 1944). Again, apart from Sri Lanka and India's east coast, there seem to be no examples of flat-bottomed plank boats in the region. The only exception may be the *ramas* of Sudan, a small, simple, fine-sterned craft used as a dhow tender, which seems to have been developed because of a shortage of logboats from Kerala!

Thus it is perhaps unsurprising that nowhere else in the Indian Ocean region has a craft been developed which is similar to the *paruwa*. Only Sri Lanka appears to have the necessary scow-ended logboat building tradition. The lack of such a tradition may also explain why in China, where many boats are flat-bottomed (and some even have a marked tumblehome), there appears to be no evidence of chine strakes.

179

Yet there are anomalies. Not all Sinhalese craft can be argued to be logboat-derived. Until the early years of the twentieth century a large sailing vessel – the *yathra dhoni* – was produced on the island. The account of it given by Hornell (1943: 43–6) does not note any features that can be associated with the *oruwa* except its sewing pattern and the presence of an *oruwa*-style outrigger. In fact he states that '. . . the design approximated closely to that of the Coromandel *masula* boats' (Hornell, 1943: 44). And where did the *teppam*, the log raft, come from?

Although the boats of Sri Lanka share with several other boat types of the Indian Ocean a common technique of fastening planks, indeed a specific method of sewing, this is a single attribute and not sufficient to place Sri Lanka within a broad 'Indian Ocean boatbuilding culture'. In terms of maritime ethnotechnology, Sri Lanka has a quite distinctive culture: sewing may be the only imported trait (although it cannot be ruled out that it developed here first).

However, it has been argued here that the method of building a *paruwa* is derived from the method of building an *oruwa*: that is, a boat with chine strakes has developed from a logboat building culture. The international aspects of this hypothesis are considered in the Annex to this chapter.

Acknowledgements

The fieldwork on which this chapter is based was funded by the National Maritime Museum, Greenwich. I wish to thank R. Gunaratne, W.P. Mahendra and E.C. Tranchelle for their help in the field.

ANNEX: OTHER BOATS WITH
HOLLOWED CHINE STRAKES

Oceania and Central and South America are the primary areas for logboat building today. In the former, however, there seems to be no evidence of boats with chine strakes (*iles*, transition strakes). Possibly multi-hulls or platforms built on outrigger booms fulfil any needs for extra carrying capacity. But in Central America chine strakes still exist (Roberts and Shackleton, 1983: 31):

> ... a cargo boat used on the Mosquito coast rivers was the *bateau*, built by cutting a large pitpan [a punt-ended flat-bottomed logboat] lengthwise into two equal halves. Flat bottom boards were inserted, the two halves rejoined and then further planks used to raise the sides. The average *bateau* had six paddles and carried $2\frac{1}{2}$ tons. Others, up to 6 feet in length, were capable of freight loads up to 5 tons.

Although not illustrated, it seems likely from this description that the *bateau* would have been scow-ended, like the *paruwa*. It is interesting that both these craft are derived from boatbuilding traditions that produce flat-bottomed and punt-ended logboats. However, Roberts and Shackleton (1983: 32) also publish a photograph of a dory on the Belize coast which appears to have a central plank inserted between two logboat halves with a washstrake added. Although this does appear to be flat-bottomed, it is fine-ended. It must also be noted that although flat-bottomed, punt-ended logboats are common in South America, no craft with chine strakes has been recorded there.

In Europe, perhaps only in one area of Poland does a logboat building tradition survive today. Litwin (1985: 328, 337, fig. 23.2) notes a single example, with a unique history, from the San River of an unextended logboat which was sawn in half longitudinally and a central plank then inserted. Although he illustrates plank boats modelled on logboats and notes the continued existence of logboats on the River Bug, no current Polish boats seem to have chine strakes.

However, historically (up to the nineteenth century) a rich area for boats with chine strakes was the River Adour complex in southwest France. Beaudouin (1970: 76–81) describes two logboats in the area – fine-ended and flat-bottomed with tumblehome – and then goes on to describe no less than six types and a further two subtypes of craft with chine strakes. Although all are fine-bowed, only two were also fine-sterned: one was transom-sterned and the remainder were scow-sterned. Of equal interest is the cross-sectional shape of the chine strake on these boats. Whereas those of the *paru* are all variations on a basic 'C' shape, the Adour boats have six different cross-sectional forms (Fig. 6.9).

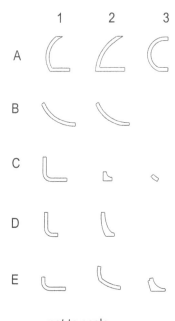

not to scale

Figure 6.9 Chine strakes – types and variations.

The ethnographic evidence presented here suggests that, although they are uncommon, boats with chine strakes are most likely to be found where:

1. There is an active logboat building tradition;
2. The logboats are flat-bottomed;
3. There is a requirement for more carrying capacity than the standard logboat can provide.

The phrase 'most likely' is used because there is at least one case where it does not apply. This concerns C3 in Figure 6.9, the most minimal of chine strakes. It is found in Venice on the *sandolo puparin* and the gondola (Pergolis and Pizzarello, 1981: 62, 78) but is not present on any other craft of the city or lagoon. It seems unlikely to be a coincidence that the chine strake (*copo*) is found on the only two craft of Venice that are built asymmetrically. It is probable that the *copo* provides some additional longitudinal strength, but this leaves open the questions: did all Venetian craft once have chine strakes, and are the *sandolo puparin* and gondola the only survivors of a change of technique, or is the *copo* a specific solution to a problem in asymmetrical boat construction? There is also the anomaly that, unlike in Sri Lanka, Nicaragua and the Adour region, there are no logboats in Venice today.

Ellmers (1984: 155), Lehmann (1978: 259) and De Weerd (1978: 15) have all characterised the chine strake as a half-logboat, implying that in order to construct a boat with chine strakes the first step is to split a logboat. While this may be true of the *bateau*'s chine strakes, it is not the case of those on the modern-day *paruwa*. If put together, the *paruwa*'s chine strakes would not make a logboat but a closed hollow log. Of all the chine strakes classified in Figure 6.9, only B1, B2, C1 and E2 would make credible half-logboats. What is argued here is the important point that the chine strake of the *paruwa* is the product of a logboat building mentality.

Because both the chine strake and the logboat are carved from a log, it is difficult to resist imposing a relationship. In some cases however the chine strake may be no more than an individual solution to the problem of providing a strong and watertight joint at the chine, completely unrelated to logboat building. However, if the chine strake were in each case a technique unconnected with logboats, why did it vanish so completely along the Rhine, where ancient logboats have been excavated, even though flat-bottomed boats persist to this day (Keweloh, 1985)? It would seem that a sophisticated practice gave way to much simpler techniques, principally fastening the outboard edge of the outer bottom strake to the face of the lowest side strake – what Ellmers (1984: 163) terms a 'composite chine girder':

> Apart from its integration as a longitudinal element in planked boats the chine girder itself is a survival from the phase of mere logboats. That is the reason why the boatwrights tried to eliminate it and to substitute composite chine girders which were much more in line with the techniques of plank boats.

This is not very convincing. It overlooks the points that the chine strake is time-consuming and difficult to make and therefore expensive. It is possible that once logboats ceased to be a major form of water transport the technique of hollowing was abandoned or forgotten. We tend to look for developments in boats and often overlook degeneration or simplification. The *madel paruwa* of Pâris's time had grooves cut to protect the sewing – a sensible procedure, but no longer practised. This is degeneration. Perhaps the sand *paru* observed without a chine strake but with an external batten instead are another example of degeneration. They may herald the beginning of the disappearance of the chine strake in Sri Lanka.

183

7

THE *VATTAI* FISHING BOAT AND RELATED FRAME-FIRST VESSELS OF TAMIL NADU

Seán McGrail, Lucy Blue, Eric Kentley and Colin Palmer

During an exploratory visit to Tamil Nadu in 1994, vessels were seen being designed and built by 'frame-first' methods at a ship-building site in Tuticorin and at a boatyard in Atirampattinam (Fig. 7.1). These vessels were different in this fundamental aspect from all other traditional types of boat on the Bay of Bengal coast, which are built 'plank first'. It was clear that these Tuticorin and Atirampattinam boats and ships needed further study, for they had never been recorded in detail, and they were increasingly being fitted with engines. Furthermore, it seemed likely that they might also shed light on a period of technological change in late medieval Europe when plank-first methods were increasingly replaced by frame-first methods.

Subsequently fieldwork was undertaken during January 1997 and February 1999 with the aim of recording Tamil Nadu frame-first techniques, in particular the methods of design.

The coastal region of southern Tamil Nadu

Geomorphology and weather

The coast of Tamil Nadu (see Fig. 7.1) includes the southwestern reaches of the Bay of Bengal (the Coromandel coast), Palk Bay and the western shores of the Gulf of Mannar. This coastal region is generally low and sandy, with an almost continuous line of sand dunes that run parallel to the shore. The dunes are intermittently backed by elongated tidal lagoons with marshy tracts on their landward side, some of which have been silted by aeolian and alluvial deposition (Ahmad, 1972; Srinivasan and Srinivasan, 1990; Kalyanasundaram, 1943). The southern part of this coast is generally classified as an emergent or advancing type (Loveson, Rajamanickam and Chandrasekar, 1990: 161), with sections of the coastline subject to progradation where the alluvial

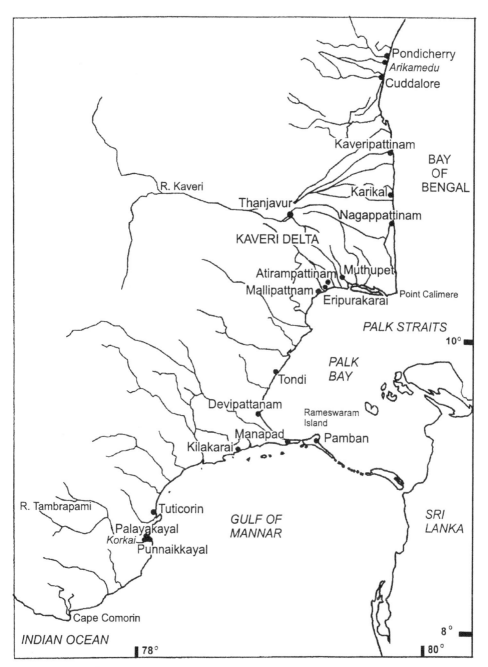

Figure 7.1 Map of coastal Tamil Nadu showing sites mentioned in the text. Data taken from *Atlas of India* (1990: Oxford) and Admiralty Chart 69.

deposition from river systems is intense. For example, in the Kaveri (Cauvery) delta it has been suggested that progradation probably amounts to 5 m per year (Bird and Schwartz, 1985).

Ancient beach ridges have been observed all along the coast, indicating a series of two to eight sea-level stands, some of which extend as far as 25 km inland (Loveson, Rajamanickam and Anbarasu, 1990). Over time, many of these ridges have been affected by intensive marine action, alluvial deposition and, more recently, coastal uplift which is particularly noticeable to the north and south of Rameswaram Island (Bird and Schwartz, 1985).

Tidal ranges (*c.* 1 m) off the Tamil Nadu coast are not as large as along the northern shores of the Bay of Bengal and the coastal shelf is generally shallow, gently tilting towards the sea. Maritime conditions in the region are dominated by monsoon winds: the SW monsoon from mid-April/early May to mid-August/September; the NE monsoon from mid-October to January. Outside these seasons, diurnal land and sea breezes usually prevail, while conditions in March and April are generally calm (*Bay of Bengal Pilot*, 1887).

Much of the coastal littoral has been formed by the action of monsoon winds which, in places such as the southern reaches of the Gulf of Mannar (Ahmad, 1972), run parallel to the shore, leading to longshore drift and the formation of barriers, lagoons and spits. The most pronounced effects of longshore drift may be seen in the region of Rameswaram (Pamban) Island where a low, sandy spit, some 22 km in length, marks the northern reaches of the Gulf of Mannar. This spit is the westernmost of a chain of islands and submerged shoals, formerly linked by Adam's Bridge ('Rama's causeway'), which extends from Sri Lanka to the Tamil Nadu coast. This natural bridge seems to have existed until the late fifteenth century AD when an earthquake divided the isthmus into several segments, leaving a series of reefs in shallow waters.

Early ports

The coastal waters of Tamil Nadu (see Fig. 7.1) generally offer little shelter to vessels, especially during the NE monsoon, except in the region of Tuticorin where there is a long chain of detached beds stretching from Manapad to Kilakarai which runs parallel to the coast and protects inshore waters (Hornell, 1945: 214). Tuticorin itself was an important landing place from the mid-sixteenth century but was not developed as a port until the mid-nineteenth century (Deloche, 1994: 95). Elsewhere in early times, shipping sought shelter behind barrier bars and in deltas such as that of the River Kaveri, to the north of Palk Bay, and the Tambraparni delta in the Gulf of Mannar. Shallows and progradation of the coast are more prominent in deltaic regions where alluvial deposition and the shifting of former river beds have cut off a number of the former river ports once located in river mouths.

An example of such a port is Korkai which, *c.* 2000 years ago, was a river anchorage in the River Tambraparni: it now lies some 8 km inland, buried beneath alluvial deposits. The site is mentioned in the *Periplus* and has been identified with the ancient provincial capital of the Pandyas region; it is probably the emporion known to Greek geographers as Kolkhoi.

When Korkai decayed as a result of siltation of the river mouth, it was replaced as an emporion by Kayal (Cail), described in AD 1292 by Marco Polo as a great and noble city, a seaport and a centre of the pearl fishery that was visited by many ships, some from as far away as Hormuz, the Persian Gulf, Aden and Arabia (Caldwell, 1881: 40). Kayal was originally at the mouth of the River Tambraparni (Hornell, 1922: 8–9), but Caldwell (1881: 37) identified it with the hamlet of Palayakayal, which is now *c.* 3 km inland (4 km downstream of Korkai). Remains covering an area more than 3–5 km in radius suggest that Kayal had fallen into ruin by the mid-sixteenth century. At about that time the Portuguese established a third port, Punnaikkayal (new Kayal), at the contemporary mouth of the river, 4 km SE of Palayakayal.

Kaveripattinam, a large Chola port on the Coromandel coast, was known to the Greeks as Kamara and to the Romans as Khaberis. In Tamil literature it is called Puhar, where there was a colony of Roman merchants (Casson, 1989: 25). Siltation of the Kaveri delta and encroachment of the sea probably made this port obsolete by the fifteenth century AD, and today the river cannot even traverse the offshore bar (Deloche, 1994: 97). The site was surveyed in the 1960s and a brick structure was later identified as a jetty dated to 300 BC (Wheeler *et al.*, 1946: 19–20; Rao 1970: 94). 'At the zenith of its prosperity this city extended over an area of 9 square kilometres, and could boast a number of wharfs and warehouses built to handle cargo from distant lands' (Rao, 1970: 94).

The port of Nagapattinam, formerly the principal seaport of Tanjore (Thanjavur) District, is at the mouth of the River Uppanar within the Kaveri delta. It is assumed to be Ptolemy's Nikama and the Mali Fatan of the Arabian geographers. During recent excavations brickwork structures of the fifth century AD have been uncovered which are believed to be the remains of a wharf (Rao, 1970: 94; Deloche, 1994: 99, note 19). Nagappattinam was one of the first and main Portuguese settlements on the Coromandel coast during the sixteenth century. In 1660 it was taken by the Dutch and in 1681 by the English (Deloche, 1994: 99). During this colonial period, it was one of a number of Coromandel ports that were used by vessels of up to 200–300 tonnes; they all gradually fell into disuse as flooding and siltation made them obsolete (Deloche 1985, 1994: 99; Thivakaran and Rajamanickam, 1994).

Archaeological investigations at Arikamedu, *c.* 4 km south of Pondicherry, have identified it as the site of an Indo-Roman port of the first century AD, probably the Poduke of the *Periplus* and the emporion known to the Greeks as Podovake (Wheeler *et al.*, 1946; Begley, 1983,

1996). Other Tamil ports and landing places may be recognised in placename suffixes: *pattinam*; *pakkam*; *puram* (Rajamanickam and Jayakumar, 1991: 5).

Palk Bay

Palk Bay extends from Point Calimere in the northeast to the island of Rameswaram to the south (see Fig. 7.1). The bay is relatively sheltered and the prevailing winds and NE monsoon run roughly parallel to the west shore, forming a more or less continuous line of sand dunes some 2-5 m in height. Shallow waters, less than 3 fathoms (5 m) in depth, extend out to 8 km (*Bay of Bengal Pilot*, 1887). The southern arms of the Kaveri (Cauvery) estuary form a marshy zone along the northern shores of the bay, with extensive mud and sand banks, some 7–9 km wide and *c.* 45 km long, which are separated from the shoreline by a barrier around 25 km in length (Ahmad, 1972: 89). This barrier is generally marked by mud flats, mangrove and salt swamps, but near its southern end it is covered in sand dunes and ridges.

The low-energy coastal environment of the northern reaches of Palk Bay, the seaward migration of the shoreline, deposition of tidal sediments and fast denudational activities have all gradually contributed to the siltation of these shallow waters (Loveson, Rajamanickam and Chandrasekar, 1990: 168, fig. 6).

The ports of this region are well protected from monsoon winds, particularly those along the northern shores of the bay, i.e. the Thanjavur coast. Between Pamban and Atirampattinam a long, smooth shore sweeps towards the NE with reasonable anchorages, especially at the small open roadsteads of Devipattanam and Tondi, both formerly populous Tamil trading centres (*Bay of Bengal Pilot*, 1887). River-mouth anchorages in the southern reaches of the River Kaveri include Kottaipatam, Mallipattnam, Eripurakarai and Atirampattinam. Atirampattinam, around 48 km west of Point Calimere, is especially difficult of access and, since the late nineteenth century and probably earlier, boats have had to be manhandled through the shallows onto a muddy foreshore. Conditions are better at Eripurakarai, nevertheless boats still have to be dragged out of, and into, the sea across a wide expanse of mobile mud. Muthupet, 32 km west of Point Calimere, was also a landing place in the early twentieth century (Hornell, 1945: 222–4). A few fishing boats are still based there. However, owing to siltation at the river mouth it now lies some 10 km from the sea, a vivid reminder of the rapid progradation experienced on this deltaic coastline during the past 80 years. The fishermen of Atirampattinam and Eripurakarai continue to beach their boats along coastal sandbanks with increasing difficulty and at ever-increasing distances from their villages. Unless villages and landing places are established nearer the sea, it is conceivable that, during the course of the twenty-first century, these fishermen will lose their livelihood.

Since the mid-nineteenth century, the increasingly shallow waters in this region have forced ships to anchor offshore, and lighters have had to be used to carry goods and people between ship and shore (Hornell, 1910: 44–6). Hornell believed that it was at this time that flat-bottomed boats were evolved to undertake this task. Whenever and by whatever means the flat-bottomed, light-drafted, keel-less, planked boats that are used in this region today came into being, there is no doubt that they admirably suit the shallow, sheltered waters of Palk Bay with its smooth bathymetric environment and its estuaries, backwaters and lagoons (Thivakaran and Rajamanickam, 1994).

Early accounts of Tamil maritime activity

Classical geographers such as Strabo and Ptolemy and the author of the first-century AD *Periplus of the Erythrean Sea* refer to maritime contacts, predominantly mercantile, between the Tamil coastal region and the West and the East. Recent archaeological investigations at such sites as the first-century AD Indo-Roman port of Arikamedu have shed further light on this early maritime activity (Wheeler *et al.*, 1946; Begley, 1983, 1996).

There are also references to Tamil ventures within the Indian trading system in Sangam literature of the third century BC to fourth century AD (Hornell, 1922: 7; Thivakaran and Rajamanickam, 1994). These include references to the *paradvar* people who were fishermen before they took up trading, pearl fishing and salt production along this coast (Ray, 1994: 4, 181). Their descendants, known as *paravas*, were encountered by the Portuguese in the early sixteenth century: Francis Xavier SJ noted that, south of Manapad, these *paravas* only caught fish, while further north they also went diving for pearls and chank seasonally (Schurhammer, 1977: 306–328). The *parawars* of the late nineteenth/early twentieth century also engaged in fishing and in pearl and chank diving and, indeed, they do so today (Hornell, 1910: 50; 1922: 6–20; Caldwell, 1881: 37, 80–82; Ray, 1994: 14). The Portuguese first settled in Tuticorin in 1543, followed by the Dutch in 1658 and finally by the British in 1796, and during this period Tuticorin became the focus for the pearl fishery.

Early descriptions of boats and ships

Among vessels mentioned in early Tamil literature is the *toni*, which was evidently a logboat, and this term continues to be used today when referring to such boats. However, the term *toni* is also used today for bigger, planked vessels (Arulraj and Rajamanickam, 1988: 11).

One of the earliest Europeans to mention the fishing vessels and coastal traders of Tamil Nadu was the Portuguese missionary Francis Xavier SJ (Schurhammer, 1977: 293–328). In 1542–3 he noted that, in southern Tamil Nadu, the *paravas* had large open boats, known as *vallam* and *toni*, with mast, sail and oar. Both types were used for net fishing; the *toni* was also used for coastal voyages between settlements, and some went on trading voyages to Sri Lanka and even as far as the Maldives. Fishing nets were of coconut fibres and had wooden floats and stone sinkers. Sails were made from cotton, dyed russet and strengthened by being boiled in water to which cow dung and selected roots had been added. Details of the structure and shape of these vessels were not noted in these early accounts, however, and so it is not possible to determine now whether there has been any continuity in building techniques between the sixteenth-century *vallam* and *toni* and those of the twentieth century. There does, however, seem to be some continuity in the uses to which these vessels were put.

In a series of papers based on fieldwork in the early twentieth century, Hornell (1910, 1916, 1920, 1945) discussed some of the Tamil coastal craft. Three of the craft mentioned by him are relevant to the present enquiry:

a) The *ballam* or *vallam* (Hornell, 1920: 155–6, fig. 4; 1945: 216–17, plate 1A).

Most of these were expanded and extended logboats with rounded bottom and flared sides, up to 30 ft (9 m) in length and with a single lugsail. Larger ones, measuring approx. 34 x $6\frac{1}{4}$ x 2 ft (10 x 2 x 0.8 m) and of two tons cargo capacity, were solely plank built: they had a second lugsail on a mizzen mast. It is not impossible that there is a connection between the plank version of this *vallam* and the *vallam* of today, but Hornell does not give sufficient structural information for this hypothesis to be tested.

b) The Tuticorin cargo lighter or *dhoni* (Hornell, 1910: 50–51; 1945: 217–19, fig. 1).

This type of craft (of 15 to 60 register tons) was used in the pearl fisheries and also to lighter goods from and to ships anchored off Tuticorn. When on coastal voyages, this *dhoni* had three masts; otherwise one. Hornell (1945: 217) believed that it was of 'comparatively recent origin' and had been copied from the Arab *bum* in the mid-nineteenth century. He dismissed out of hand the possibility that this lighter had 'evolved directly from the plank-built canoe used in the coral trade' (i.e. the *vallam*). However, in shape, structure and rig the *dhoni* lighter depicted by Hornell does seem to be generally similar to his *ballam/vallam*, and it could well be a related, indigenous vessel. Whether there is a direct link between the 1914 *dhoni* and the *dhoni/thoni* of today is now impossible to say; nevertheless, it may be that, in general terms, the latter is a larger version of the former.

A photograph, probably taken in the 1940s, of several *toni* in Tuticorin harbour (Fig. 7.2) was published by Brodrick (1952: 161) in a biography of Francis Xavier. These ships appear to be generally similar to the *toni* noted by

Figure 7.2 Thoni in Tuticorin harbour, sometime before 1950. From Brodrick (1952: 161).

Hornell in the first decade of the twentieth century (1945: 217–19, fig. 1), being double-ended with little sheer and having a single mast forward of the hold amidships. They were probably at the upper end of Hornell's size range.

c)   The *vattai* or *vattal* of the northern coast of Palk Bay (Hornell, 1920: 162–6; 1945: 222–7, plates 2B, 3A, figs 4, 5 and 6).

Hornell called these boats '*vala vattai*' (1945: 223) and 'palagai kattu *vattai* or *vattal*' (1945: 226), and he considered that the term *vattal* could be derived from the Portuguese *batel*, meaning 'small boat'. These vessels were to be found at Muthupet (Fig. 7.3) and Atirampattinam (Fig. 7.4) and were used for fishing in Palk Bay. They were either extended logboats or narrow 'carvel-built' plank boats, measuring some 43 x $4\frac{1}{2}$ x $2\frac{1}{2}$ ft (13 x 1.4 x 0.76 m). They had up to three masts, each with a single lugsail, and, under full sail in a good breeze, they were 'accounted the swiftest in these seas'.

Hornell noted that these vessels had up to three leeboards and one, sometimes two, balance boards, the longer ones being made in three sections. These boats generally had a five-man crew. Hornell also noted (1945: 224) that the Muthupet boats had a large rudder hung by gudgeon and pintle at the lower end and by a coir lashing below the tiller (Fig. 7.3). Yet, at the same time, he claimed that boats from Atirampattinam (only a few miles west of Muthupet – see Fig. 7.1) did not use a fixed rudder but had 'a pair of

191

Figure 7.3 A *vattai* at Muthupet, drawn by James Hornell in 1914. After Hornell, 1945: fig. 4.

quarter steering boards' having 'the shape of local lee boards' (Fig. 7.4). This appears to be an example of Hornell's over-hasty interpretation of his observations. It seems very likely that what he had seen was the temporary use of a leeboard fitted with a simple tiller to steer a *vattai/vattal* when sailing stern first, as they almost invariably do when leaving their mud berths at Eripurakari today (see page 226).

Most of the *vattai/vattal* features described by Hornell, including 'carvel-built' (i.e. frame-first), are found in the *vallam* today; and the one principal structural difference with today's *vattai* is that the latter has a markedly rising sheerline towards the bows.

Hornell's remarks about all three of these types seem to suggest that he encountered, but did not fully recognise, the use of several names for one type of vessel and the use of a single name for several different types. Similar inconsistencies in local terminologies have been noted by Deloche (1994: 156) in relation to the boats of the Ganges. It seems likely that Hornell's plank-built *ballam/vallam*, *toni* and *vattai/vattal* were, in essentials, all very similar and sufficiently alike that today they would be considered variant forms of one tradition of ship and boat building.

In the mid-1940s Hasler (1950) saw what were known as *vallam* at Kankesanturai on the north coast of Sri Lanka. These boats, which had been built in India, had a balance board and similar planking to those Hornell

Figure 7.4 A *vattai* at Atirampattinam, drawn by James Hornell in 1914. After Hornell, 1945: fig. 6.

(1945: plate 2B, fig. 4) had seen at Muthupet in 1914 (Fig. 7.3). Hasler noted that these boats had fine sailing qualities and that the newer and faster ones had a blunt 'cruiser' stern with a very flat run and a hard turn of bilge, rather than the sharp canoe stern of the older boats.

Hawkins (1965) discussed aspects of the Tuticorin *thoni*'s structure and use in an article based on fieldwork in the 1950s. He also included the Tuticorin *thoni* and the Cuddalore dinghy in a book on sailing ships of the Indian Ocean (1977). He considered that the *thoni* was 'a comparatively new arrival' and quoted Hornell (1945) for its 'evolution' during the early years of the twentieth century. His observations on the building of a mid-twentieth-century *thoni* are, on the other hand, based on his own fieldwork.

Hawkins noted that Tuticorin *thoni* were double-ended with a hard chine, practically no sheer and no bulwarks. Their overall shape was devised to give maximum cargo capacity. They generally had three masts with lateen sails on fore and main and a gaffsail on the mizzen (Fig. 7.5). There were two jibs forward, a square topsail could be set on the foremast, and a spritsail was set

193

Figure 7.5 The *thoni* 'Ave Maria' in Tuticorin *c.* 1960. From Hawkins, 1965, plate 10.

from the bowsprit. *Thoni* were steered by tiller rather than a wheel. In favourable conditions they could achieve 12 knots.

In the 1950s two or three were generally built each year. Between 1958 and 1961, however, there was an average of ten a year: some of these had bigger hulls, a fourth mast and a crew of up to seventeen. In 1960, 125 *thoni* were registered at Tuticorin and 10 at nearby Kulasekaripatinam. They were principally involved in trade between Tamil Nadu, Sri Lanka and the Malabar coast.

Hawkins (1965) emphasised that *thoni* were not built from plans and mentioned that the outlines of the frames were marked out full size on a floor, but he did not go into details of the design methods. The frames (*karavai*) were made of *babal* (if not available, then *portia* or *reem*); *venteak* was used for the planking, *karumarudu* for keel and posts, and *punnai* for masts and spars.

A.W. Kinghorn, a master mariner, has recently (1996) published a description of the sailing rig of Tuticorin *thoni* he had seen in Colombo harbour in 1994 when there were still some purely sailing craft, although by then most had an auxiliary engine installed. The sailing vessels were double-ended, without bulwarks and superstructure, and were manned by a crew of ten or so. The larger vessels had three masts and a long bowsprit: the foremast had a lateen topsail as well as a mainsail, the mainmast a single lateen, the mizzen a gaffsail; while there was a lateen jibsail between foremast and bowsprit. Bonnets, staysails and 'lateen stu'nsails' could also be set (Fig. 7.6).

Figure 7.6 A *thoni* under all plain sail in Colombo, 1994. Photo: Captain A.W. Kinghorn.

Fieldwork

During January 1997, working under the aegis of Tamil University, Thanjavur, we visited boatyards at Atirampattinam, shipbuilding sites at Tuticorin and Cuddalore, and beach landing places at Atirampattinam and nearby Eripurakarai, as well as others near Cuddalore (see Fig. 7.1). The Tuticorin *thoni* (Fig. 7.7) and the Cuddalore *kotia* (Fig. 7.8) were examined and photographed while they were being built on foreshore building sites,

Figure 7.7 A *thoni* in frame on a foreshore building site at Tuticorin.

195

Figure 7.8 A *kotia* at Cuddalore nearing completion.

and the *mistri* – the master shipwright in charge of each of these sites – was asked about his design methods and building techniques. The *vattal* dumb lighter and the *vallam* fishing boat were noted but not recorded further, as time did not allow this. The *vattai* fishing boat, on the other hand, was documented in detail: while being built at Dawood's yard at Atirampattinam, while beached at Eripurakarai (Fig. 7.9) and other nearby sites, and while underway in the northern coastal waters of Palk Bay.

A further season of fieldwork was undertaken during January 1999, aiming at interviewing a wider range of builders and fishermen and noting any variations from the design, building and operational practices already documented. The sites visited included: the one *thoni* shipyard active at that time in Tuticorin; *vallam* boatyards in Tuticorin and further north; and *vattai* yards on the western and northern shores of Palk Bay. Boat operations were observed and fishermen interviewed at a number of boat landing places in this region.

Type names

The vessels found at these sites were known to their builders and to their crew by the type names generally used in this chapter. Earlier investigators, such as Hornell (1945) and Rajamanickam and Jayakumar (1991), sometimes used other names to describe vessels which appear to be similar to the ones described here. In particular, the latter authors (1991: 15, 16, 19,

Figure 7.9 A *vattai*, under foresail and mainsail, taking the ground at the beach landing place at Eripurakarai.

112, 113; figs 8, 9; plates 4d, 5b, c, d and 11a) used the term *vatta* for the vessel we call *vallam* and *vallam* for the boat we call *vattai*. The main features of these ships and boats are described below using our preferred type names with alternative spellings in parentheses.

Thoni (toni)

A motorised sailing ship with three masts built in Tuticorin for trading with Sri Lanka. These ships have a full form, with a near-rectangular cross-section, and they range in capacity from 225 to 650 tonnes, with keel lengths from 29 to 38 m. Their length/beam/depth ratio is 8:2:1.25. Nowadays engines are used when leaving and entering harbour; once clear of land, sails are set.

Kotia (kottiya)

Built in Cuddalore principally for the west-coast trade. They are a smaller version of the *thoni*, with only one mast.

Vattai

A flat-bottomed fishing boat found at several sites in Palk Bay where it seems to predominate. In size they range from *c.* 13.72 m long, with a beam of 2.13 m and a depth of 1.37 m, to the smallest vessel noted of *c.* 5.18 x 1.07 x 0.76 m. The one chosen to be documented was 12.09 x 1.60 x 0.93 m. Irrespective of size, the larger ones are all similar in shape with marked sheer, especially forward. They have two or three masts, each with a lateen sail. As they are flat-bottomed and high-sided, this large sail area would give them excessive leeway (drift downwind) were it not for the use of leeboards (rare on this coast). Excessive heel is avoided by using a balance board.

The smaller ones, less than *c.* 7 m in length, have a gentle sheer, with only a slight increase in curvature towards the bow. They have a very short parallel midbody, and they are relatively beamy boats with an overall L/B ratio of between 4:1 and 5:1, compared with the 6 or 7:1 of the larger boats. They also have relatively more freeboard. They have a single mast, and a balance spar rather than a board.

More than one hundred *vattai* fishing boats sail out of Atirampattinam and the nearby landing sites of Mallipattnam and Eripurakarai. Large fleets sail out of Devipattanam and Thondi, further south in Palk Bay. It looks as though Adirampattinam is towards the northern end of the distribution, as there are only a few at nearby Muthupet and at Vederanyam. At the latter site, which is about 1 km west of Point Calimere, *vattai* were moored close to a beach with road access.

Vattal

A dumb lighter, about 15 m in length, is built and used in Cuddalore. Its basic structure is similar to that of the *vattai*, but it has an open hold with walkways either side and has only a very little sheer forward.

Vallam

An open boat, 7 to 10 m in length, nowadays almost always motorised, but often with auxiliary sail on one or two masts. They are used for offshore fishing and are found intermittently from Cuddalore southwards but are more common in the southern reaches of Palk Bay and in the Gulf of Mannar south of Tuticorin.

Structurally *vallam* are generally similar to *vattai*, except that they have a plank-keel rather than a foundation plank. Like *vattai*, *vallam* are double-ended but, in marked contrast to *vattai*, they have a gentle sheerline, with only slightly more freeboard forward than aft. Stem and stern posts are straight and have little rake; the midships region has little or no flare.

As in the *vattai*, the frames of the *vallam* are designed using the scrieve-board system (see pages 202–210). The main difference is that the *vallam* has no passive frames: all are designed. *Vallam* have only a few frames equal in shape to the master frame (e.g. 2 out of 17, i.e. 12 per cent); thus they have a relatively short parallel mid-section when compared to the box-like hull of the *vattai* (e.g. 14 out of 27, i.e. 51 per cent).

The impression gained during January 1999 was that, compared with 1997, there had been an increase in the number of *vallam* both under construction and in use. In a *vallam* yard in Tuticorin, for example, there was effectively series production. The use of sail ('sail assist') was more in evidence north of Tuticorin, with the boats at Thondi having not only two masts, each with a sail, but also balance boards and leeboards, as well as an engine.

Plank-first and frame-first methods

It will be argued later in this chapter that these seemingly disparate craft all belong to a single boatbuilding tradition, 'tradition' being defined as 'the perceived style of building generally used in a certain region during a given time range'. One feature which links these five vessel types together is that they are all built frame first.

Wooden boats and ships, worldwide, can be divided into two main groups (Fig. 7.10): those that are built plank first (otherwise known as 'shell-built'); and those that are built frame first (also known as 'skeleton-built'). The difference between these two groups lies in the way the builder conceives, visualises or designs his boat. In the first case the builder fashions the planking and fastens it together to form the shape of hull he wants; then he fastens framing inside the planking to support that planked shape. In the second case he fashions and erects the framing to form the hull shape required; and then he fastens planking to that framework to make it watertight. An ethnographer recognises which method is being used as he records a boatbuilder's sequence of actions and studies the building aids in use. An archaeologist does not have so straightforward a task when examining an excavated boat: the internal stratigraphy of the boat has to be deduced, usually in a laboratory or workshop well after the excavation, by detailed examination of planks and frames so that the sequence in which the various timbers were added to the boat's structure during the building process can be worked out.

All the earliest known planked vessels were built by plank-first methods (McGrail, 1996: 72–80). The oldest planked vessel in the world, Cheops' royal burial ship of the mid-third millennium BC; wrecks of the mid-second millennium BC excavated in the eastern Mediterranean and in the Humber estuary in Britain; the ships of the Vikings; and the earliest known Arab, Indonesian and Chinese plank boats – all these were built plank first. The majority of South Asian traditional vessels are still built plank first.

The earliest evidence for the alternative technique – the use of the frame-first approach – comes from northwest Europe of the late Roman period. Seagoing ships and boats of an indigenous building tradition, known to maritime archaeologists and historians as 'Romano-Celtic', were built frame-first during the first to fourth centuries AD (McGrail, 1995, 1997). This was

(A)

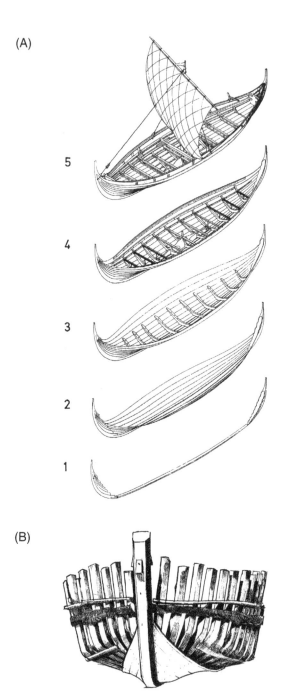

5

4

3

2

1

(B)

Figure 7.10 A: The 'plank-first' sequence of building a boat. After Crumlin-Pedersen, 1983: fig. 5. B: European twentieth-century 'frame-first' boat under construction.

a simple version of frame-first building. The framework was not fully built before planking-up began, rather an alternating procedure was used: first some framing was erected, then some planking was fastened to it, then more framing, and so on. Nevertheless, as in the fully frame-first procedures used in the eighteenth and nineteenth centuries, the framework determined the shape of the hull. Furthermore, such a framework had to be 'designed'.

Frame-first methods next become archaeologically visible in the Mediterranean, in wrecks dated from the sixth to eleventh centuries AD (McGrail, 1997: 79). Subsequently, in the fifteenth and sixteenth centuries, there was a marked shift from plank-first to frame-first methods in the central and western Mediterranean, as ships were built for exploration of the Indian Ocean and the western Atlantic. Sixteenth- and seventeenth-century texts have survived which describe, albeit not unambiguously, the design and building of ships by frame-first techniques, methods which have come to be known to scholars as 'Mediterranean Moulding' (Sarsfield, 1984; Bellabarba, 1993; Greenhill, 1995: 256–73; Rieth, 1996). Recent use of similar methods has been noted in Brazil (Sarsfield, 1988; Carrell and Keith, 1992; Barker, 1993) and in Greece, Lake Geneva and Marseilles (Rieth, 1996).

Aims and research methods

It was decided to concentrate research in 1997 on the *vattai* and to use information gleaned from the builders and users of the *thoni, vallam, vattal* and *kotia* wherever it was applicable. The aim was to compile all the data needed to build a model of a typical *vattai*, accurate in all essential details, using the building procedures found in Tamil Nadu. Numerous interviews were undertaken (with the assistance of an interpreter), construction and boat use were observed, and individual boats were documented.

Fieldwork was mainly undertaken at Atirampattinam and Eripurakarai, on the northern shores of Palk Bay, but vessels in building yards and at landing places in Tuticorin, in the Gulf of Mannar and in Cuddalore were also examined. Vessels of these five types were not found at other landing places on the Coramandel coast between Point Calimere and Cuddalore, nor further north.

At Atirampattinam we were able to observe *vattai* being built in two boatyards (Fig. 7.11), although interviews were mainly concentrated in that of Dawood. At Eripurakarai, a beached *vattai* was selected for detailed measurement. This boat was hogged and had other damage but otherwise was a representative example of the type. She was measured and drawn at a scale of 1:10, photographed from all viewpoints, and notes on her structure were compiled. The plans of this boat (Fig. 7.12) are a rectified version of the field drawing, with the distortion and damage removed and missing parts replaced. The data obtained from the beached boat were supplemented by interviews with boat users, particularly at Eripurakarai; and two short

Figure 7.11 A *vattai* being built at Atirampattinam. The next-to-top strake and an angled bow plank have been fastened to the framework.

voyages were undertaken in Palk Bay. Some days were also spent at Tuticorin in the *thoni* building yards and at Cuddalore among the *kotia* and *vallam* builders. It proved impossible to find a boat or ship at the design stage or even one in the early phases of building; for these aspects we had to rely on descriptions rather than direct observation.

In the second season, 1999, *thoni* and *vallam* yards at Tuticorin and several boatbuilding yards and landing places in Palk Bay were visited. Dawood had retired from boatbuilding in Atirampattinam, but we were able to visit Sharif's *vattai* yard nearby. The aim was to interview a wider range of builders and fishermen and record any variations from the practices we had previously documented, especially in the design methods of the *vattai* and the *vallam*.

Designing the *vattai*

The technical terms used in this and succeeding sections are illustrated in Figure 7.13: the frame stations are numbered 1 to 27 from the bow. The *vattai* is a frame-first boat and, before building can begin, that framework has to be designed. The principles of design are:

1. To design a number of frames of the same size and shape, which will form the central section of the boat. These are known by a Tamil term which translates as 'equal frames'.

202

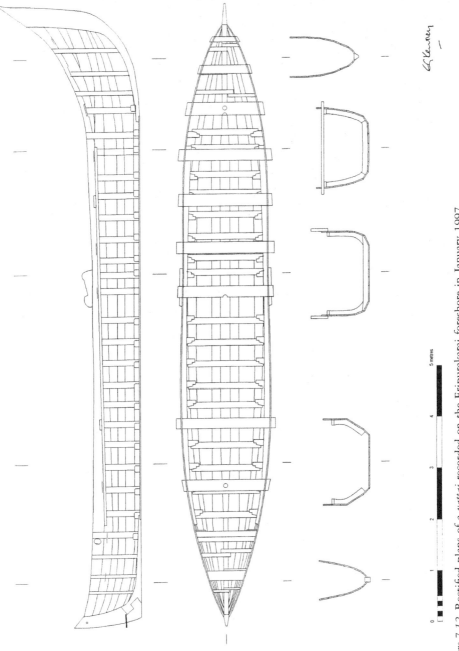

Figure 7.12 Rectified plans of a *vattai* recorded on the Eripurakarai foreshore in January 1997.

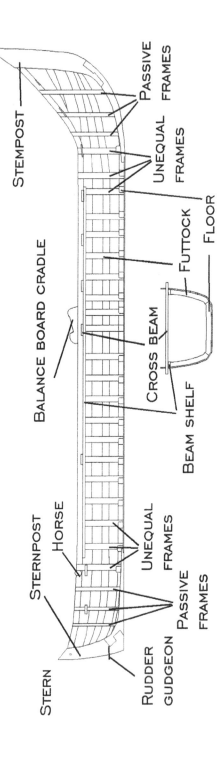

Figure 7.13 Terms used in the text to describe parts of the *vattai*.

2. To design a number of pairs of other frames, which will give the cross-section of the boat at stations forward and aft of that central section, where the hull rises and narrows: one of each pair is positioned towards the bow, the other towards the stern. These are known by the Tamil equivalent of 'unequal frames'.

Thus, in Figures 7.12 and 7.13, the 'equal frames' are the fifteen frames at stations 7 to 21: all these are identical in the size and shape of their outer faces. The pairs of 'unequal frames' are frames 6 and 22, 5 and 23, and 4 and 24. Frames within each pair are identically similar, but each pair is different from the other pairs and all are different from the equal frames.

The six other frames or, more strictly, the six pairs of half-frames – the foremost three in the bow and the aftermost three in the stern – are not part of the design process but have their shape determined by the planking (i.e. they are passive frames).

Frames were said to be spaced (centre to centre) at a standard distance apart: 25 inches (0.63 m) in ships; 16 inches (0.40 m) or 18 inches (0.45 m) in boats. Regardless of size, all *vattai* have an odd number of frames: the erection of a master frame at the centre of the boat, then the placing of one frame on either side again and again leads to this result. However, there is no fixed ratio of unequal to equal frames, and three pairs of unequal frames seem to be the usual number (see Fig. 7.13), although four were seen in one boat. To fix the shape of the hull of the *vattai*, therefore, the boatbuilder must produce a maximum of five frame shapes – one for the equal frames and up to four for the unequal frames.

The Tuticorin *thoni* and the Cuddalore *kotia* both employ this system of equal and unequal frames. For example, a 225 ton/29 m *thoni* has 17 equal frames and 10 pairs of unequal frames; while a 650 ton/38 m *thoni* has 25 equal frames and 12 pairs of unequal frames. Since the frames in these ships are spaced at *c*. 0.63 m intervals, the designed (equal and unequal) frames in the *thoni*, for example, extend for only *c*. 60 per cent of the hull length, whereas in the *vattai*, with a relatively greater length of wall-sided body and with less complex shapes at the ends, *c*. 75 per cent of the hull has designed frames. In the *vallam*, on the other hand, all frames are designed.

The scrieve board

The process of determining the shape of the *vattai* frames begins by preparing what in a European boatbuilding context would be called a 'scrieve board' (Fig. 7.14) for the boats, or a 'mould loft floor' for the ships: these are boards fastened together to form a horizontal surface on which the shapes can be drawn. The preparatory marking of board or floor is shown in Figure 7.15A. A rectangle is scribed or drawn in ink – its breadth and height being equal to the beam and depth of the intended *vattai*. These proportions are probably

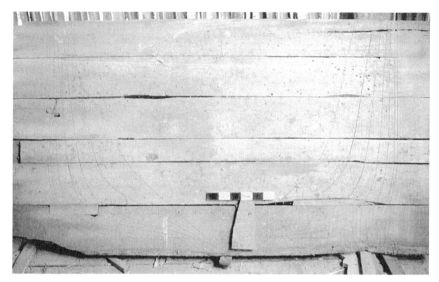

Figure 7.14 A *vattai* scrieve board used in a boatbuilding shed at Rameshwaram, with the shapes of the master frame and six unequal frames scribed by knife. Scale division 50 mm.

determined by formulae (as are those of the *thoni*), but we were unable to establish this precisely for the *vattai*. The sides of the rectangle representing the breadth of the boat are bisected and a vertical line representing the boat's middle line is drawn. In 1997, at Dawood's yard in Atirampattinam, the half-breadths were again bisected; these half and three-quarter beam positions are used when transferring a frame shape from the scrieve board to a crook (see page 211). These lines no longer appeared on any board seen in 1999, and it seems likely that they have been redundant for some time. Finally, diagonal lines are drawn between the centre of the upper side of the rectangle and the lower corners.

As can be seen from Figure 7.12, the *vattai* is wall-sided and flat-bottomed at its midship section. Therefore, as far as design of the equal frames is concerned, all the boatbuilder must do is determine the curve of the bilge. To do this he uses an *asil* – a thin wooden mould, pattern or template (Fig. 7.15B) – the shape of which is that of the master frame (which is destined to be placed close to, or at, the midship section). In height, moulds are approximately the same as the boat; in length they are greater than half the boat's breadth. Some builders have a number of moulds, each one appropriate to the construction of a particular size of *vattai*. Others mark the middle line for different breadths of boat on the horizontal arm of a single mould. The various sizes of boat designed in this latter way will have the same bilge radius; thus they will not be geosims (geometrically similar in shape) and will therefore have somewhat different hydrostatic and hydrodynamic qualities.

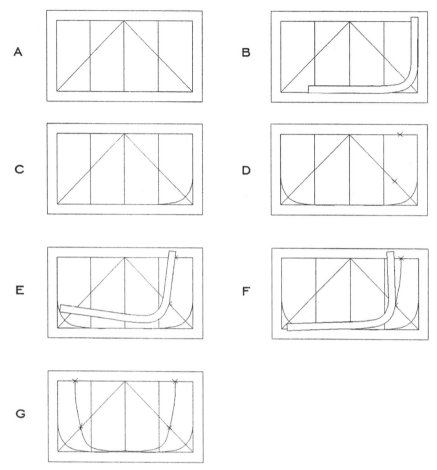

Figure 7.15 Diagram to illustrate the design of *vattai* frames using a scrieve board and a single mould: A. Preparatory marking of the scrieve board; B. Mould positioned on the board to give the shape of one half of the master frame; C. The shape of one half of the master frame (also that of all the equal frames) marked on the board; D. Points (X) marked on the board (according to a formula) to give the rising (along the diagonal) and the narrowing (along the top edge) of a pair of unequal frames; E and F. Mould positioned on the two points and then rotated to give the shape of one half of a pair of unequal frames; G. The shape of a pair of unequal frames marked on the board, alongside that of the master frame.

The mould is placed within the rectangle scribed on the board and the curve of its outer edge is marked in, from the upper horizontal line to the centre of the bottom (Fig. 7.15C). The mould is then flipped over so that a similar curve can be drawn on the other half of the board (the mould is reversible as the ends of any strengthening struts are inlaid). The shape now on the board (Fig. 7.15D) is that of equal frames 7 to 21 of the boat in Figure 7.12, including the master frame 14.

To draw the shapes of the pairs of unequal frames (for example, frames 4 and 24 in Figs 7.12 and 7.13), the boatbuilder uses a formula which is based on the total rising and narrowing of the unequal frames. The formula used by Dawood requires that, for every inch (25 mm) measured along the scrieve board diagonal from the intersection of the last drawn frame, the half-breadth narrows by $\frac{3}{4}$ inch (19 mm). Sharif's formula had less uniformity: for every inch moved along the scrieve board upper horizontal line, the diagonal measurement decreases by $1\frac{1}{2}$ inches (38 mm) for the first pair of unequal frames; by $2\frac{1}{2}$ inches (64 mm) for the second pair; and by $3\frac{1}{2}$ inches (89 mm) for the third pair. The use of differing rules means that Sharif's boats would be finer than those of Dawood.

For each pair of unequal frames, these two points (Fig. 7.15D), one on a diagonal, one on the upper horizontal line/sheerline, are marked on the scrieve board and the mould is used like a huge French curve to determine the shape of that particular pair. The mould is first used to draw a line through the two points (Fig. 7.15E). In Dawood's yard at Atirampattinam in 1997, the mould was then rotated to obtain the curve of the bilge (Fig. 7.15F). At that time the degree of rotation was unclear to us. However, at Sharif's Atirampattinam yard in 1999, this lower part of the curve was completed by eye. It may be that, in the older method used by Dawood, the degree of rotation was related to intercepts with the two (now redundant) vertical lines. The resultant composite curve (Fig. 7.15G) gives the shape of a pair of unequal frames which are both higher at the turn of the bilge and narrower overall than the master frame.

In one Tuticorin *vallam* yard, after the diagonal construction lines have been drawn on the scrieve board, a point is marked $7\frac{1}{2}$ inches (190 mm) along a diagonal from the lower right corner of the rectangle. A band-saw blade, a suitably flexible tool, is then aligned, teeth upwards, through this point, the upper right corner and the mid-point of the lower horizontal line. The blade, effectively a spline, is fixed in position by a few nails and a chalk line is drawn around the bottom of the blade. This chalk line gives the shape of the master frame, and the mould made from it is used to determine the shapes of the unequal frames in the same manner as for *vattai*.

The wooden mould for a *vallam* has (as is to be expected) more curvature in its upper parts than has a *vattai* mould. It is possible that the technique of using a band saw is also used by *vattai* builders, since their moulds sometimes have a triangular-shaped piece of wood fitted to each outer upper edge to make the edges straighter.

The sheerline

Frame shapes drawn on the *vattai* scrieve board used by Dawood in 1997 have identical depths (Fig. 7.16): that is, the sheerline is indicated as flat.

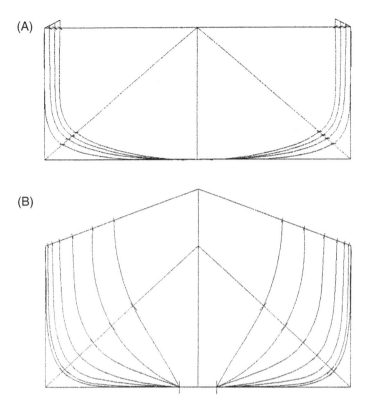

Figure 7.16 A) Diagram of a scrieve board with the shapes of a *vattai* master frame and three unequal frames marked on it. The rise in sheer is allowed for at top left and right. B) Diagram similar to A, but for a *vallam*. The rise in sheer is allowed for by additional angled lines.

The actual sheerline has a marked rise forward and a slight rise aft, and these were fixed by the height of the posts, the actual run of the sheerline being established by eye. In contrast to this, in two boatyards visited in 1999 (one *vattai*, one *vallam*), a rising sheerline was plotted on the scrieve board. For a *vattai* with three pairs of unequal frames, this line was derived by making the shape of the third unequal frame $1\frac{1}{2}$ inches (38 mm) deeper than the master frame. Using the known horizontal distance from the master frame to the third unequal frame, it was then possible to plot the heights of the other two unequal frames (Fig. 7.16A).

For *vallam*, which have no passive frames (all are designed) and a sheerline which differs from that of the *vattai*, a different procedure is used. The height of the last unequal frame is known in relation to the height of the master frame, and a line is drawn on the scrieve board connecting these two heights. On the example observed, the narrowing breadth at the various frames was marked along this angled line (Fig. 7.16B).

Standardisation

In 1997, and even more so in 1999, there was evidence for an increasing move towards standardisation, possibly engendered by the recent Tamil government-funded project to motorise *vallam*. This process has led to less reliance being placed on traditional design methods.

Once a particular design of *vattai* or *vallam* has proved itself operationally, the shapes of its master frame and its pairs of unequal frames are permanently scribed on a particular scrieve board. Some yards have taken this simplification a stage further and now make wooden moulds for each pair of unequal frames, rather than lifting off shapes from the scrieve board. One builder just to the east of Mallipattinam has taken this logic even further and uses a number of steel rods bent to the shape, not of the entire frame but, much more practically, to the shape of the individual floor timbers and futtocks.

The net result of these changes is that plotting boat-frame shapes on a scrieve board need only be done if and when a significant change in hull form is required. The use of saw mills to fashion standard sizes of floors and futtocks and to cut their complex scarf (see page 211) is a further step in the direction of standardisation of shapes and the obsolescence of the scrieve board. The recent elimination of the two vertical lines on boards (formerly used when transferring frame shapes) is further evidence for this tendency. There seems to be a similar trend in the shipyards.

Designing the *thoni*

In Tuticorin *thoni* yards, even though these ships have plank-keels and ends of more complex shape, the design system is essentially the same as that for the boats. Because of the depth and breadth of the ship's hull, and hence the size of her master frame, the *thoni* mould is in two pieces, overlapping around the bilge. We were shown the use of such moulds on a lofting floor to draw out equal and unequal frame shapes: this was generally similar to the *vattai* method. The only significant difference is that, whereas on the *vattai* scrieve board the *mistri* works from the centreline (see Figs 7.14, 7.16A), the reference for the *thoni mistri* is the keel's edge – the port edge when defining that side of each frame, the starboard edge for the other side. *Vallam* scrieve boards diagrams similarly depict the keel breadth at the centre of the lower horizontal construction line (Fig. 7.16B).

In practice it appears that this full design procedure is now seldom, if ever, used. Frame shapes for proven hull forms seem to have been preserved as standardised body plans (all the frame shapes in one diagram). Details vary from yard to yard, but one *mistri* had drawn out a body plan at a scale of 1:16 ($\frac{3}{4}$ in. to 1 ft) on cardboard. At this scale he was able to check, and adjust if necessary, the overall hull shape by eye, which would be difficult at a scale of 1:1. From this body plan he compiled a table of rescaled measurements –

that is the distances along a diagonal line and a horizontal line, as done on the *vattai* scrieve board. These measurements were then used to redraw each frame at 1:1 on the lofting floor and from these timbers were cut to shape.

Building the *vattai*

The framing

To transfer the shapes drawn on the *vattai* scrieve board to the timbers that will be converted into frames, a small stick is marked with the distances between the bottom of the rectangle and the frame shape at the half- and quarter-beam positions. This allows the mould itself to be aligned on the chosen timber in the correct position so that it can be drawn round. The quarter and three-quarter beam positions are no longer marked on the scrieve board; this alignment appears now to be done by eye.

There is some attempt to match the grain of the timber with the curve of the mould. For each frame, the builder marks three timbers: one forming the floor timber and the other two the futtocks. The floor and futtocks overlap, as can be seen in Figures 7.12 and 7.17. We were unable to establish how the length of this overlap is calculated, but it extends right through the turn of the bilge.

Once the timbers have been marked, they are transported to a saw mill where they are cut to shape. The sawn timbers vary between 100 and 150 mm in depth and breadth, but there is no structural reason for this. Both floors and futtocks are of *neem* wood (*Azadirachta indica*) although in earlier times *populina* was used. The saw mill also cuts the complicated scarfs that join floor to futtocks (Fig. 7.17) and cuts limber holes in the floors in an 'M' shape. Although we did not observe the work in the mill, the cutting of the scarfs seemed to be precise enough for the assembly of floor and futtocks to begin without further modifications. On the other hand, the futtocks often appeared to be oversized in depth (moulded) and the extra wood had to be worked away when the entire framework was faired. The use of the saw mill and other evidence suggests that, in the near future, the scrieve board may no longer be used to design the *vattai* framework.

It does not seem possible to give a name to the complex scarf between floor and futtocks, but it consists of a much-splayed dovetail and a notch (see Fig. 7.17). The same joint is found on the *vallam* and on the Tuticorin *thoni*, where it is between floor and futtocks and between futtocks and top timbers. The corresponding joints on the Cuddalore *kotia* are much simpler.

The *vattai* boatbuilder now assembles floors and futtocks to make frames. A bow drill is used to make pilot holes through the scarf joints, into which are driven 150 mm zinc-coated nails. The points of these nails are turned through 180 degrees and driven back into the wood, giving the appearance of a large staple.

211

Figure 7.17 The scarf between a *vattai* floor timber and its futtock.

The plank – a single piece of teak (*Tectona grandis*) – which will become the central bottom plank ('foundation plank') is now laid on stocks in a level position. In thickness it should be $\frac{3}{4}$ inch, but the accuracy of saw-mill cutting results in planking anywhere between 19 and 23 mm. Any fairing needed to ensure a good butt with the next plank when it is fitted is done with a chisel by eye. In breadth this central bottom plank is approximately 135 mm. On its upper face, the boatbuilder marks with a charcoal stick the position of the centre of each of the equal and unequal frames – usually 18 inches (0.46 m) apart.

The boatbuilder now places the master frame – the fourteenth – squarely in position on the foundation plank. When in place, its futtocks are forward of its floor. The builder drives nails from outboard through keel plank and floor timber and clench-fastens each nail inboard by hooking its tip back into the frame. Cotton may be wrapped around the top of the nail's shank as a form of caulking. The remainder of the equal frames are now erected, in no particular sequence, but working towards the ends: that is, frame 13 or 15 will be erected next. Frames forward of the master frame are set up so that their futtocks are aft of their floor; frames aft have futtocks forward of floors.

When all the equal frames have been erected, the boatbuilder erects the unequal frames. The relative positions of the futtock and floors follow the same principle as the equal frames, but the unequal frames are erected alternately: for example, frame 6, then 22, followed by 5, then 23, then 4 and finally 24.

Depending on the length of the central bottom plank and frame spacing, it may be necessary for the boatbuilder to erect the stem and stern posts before erecting the unequal frames – as was the case in the boat recorded (Fig. 7.12). These posts are of *vakai* (*Albizzia cebbeca*) and are of composite construction. When fastened to the central bottom plank, the posts are shored in position and they are carefully rabbeted (i.e. a groove or channel is worked in them: see Glossary) to take the squared ends of the planking. We were unable to establish whether or not there was a formula for the height of the *vattai* posts (which determine the sheer towards the ends): our principal informant assured us that the heights were known from 'experience'. However, the rake of the *thoni* posts is known in terms of a formula which relates horizontal to vertical measurements at intervals along the length of each post.

At this stage, with all equal and unequal frames and the posts erected, a strake (the next-to-top strake for most of the length of the boat) is nailed to the frames and to the stern post. At the bow, around the position of the fourth and fifth frames, a plank is scarfed in at an angle to this strake and extends to the stem post (Fig. 7.11). The function of the strake and the plank is to stabilise the framework and to act as a ribband (see Glossary) so that the shape of the remaining frames may be determined. In the example depicted in Figure 7.13, the remaining frames are 1, 2, 3, 25, 26 and 27. These 'passive' frames (which reinforce rather than determine the shape of the boat) are not true frames but pairs of overlapping half-frames, again of *neem* but of smaller dimensions. On frames 1, 2 and 3 the port half-frame is aft of the starboard half-frame and, unlike the latter, it crosses over the post. On frames 25, 26 and 27, these relative position are reversed. At this stage, the futtock heads of all frames extend above the positioned strake and subsequently have to be trimmed to size.

The planking

With all frames now in position, full planking-up can begin. The planking is of the same thickness as the foundation plank and again of teak. Planking begins with the first strake (that is the strake next to the foundation plank) and carries on upwards, on both sides, generally to the already fastened penultimate strake. Planks within strakes are not scarfed together but plain-butted at frames.

Vattai strakes, except the washstrake, are generally fastened to frames by nails driven from outboard and clenched inboard by hooking the point back into the frame, although occasional nails are clenched by turning the point through only 90 degrees so that they lie along the inboard face of the frame.

On the *vattai* measured at Eripurakarai, which had evidently seen much use, many of the nails had been driven from inboard: we conclude that this was the result of extensive repairs. Bolts are used as fastenings throughout the *thoni* and *kotia*; there is a move towards this usage in the *vattai*. During 1999, occasional use was also noted of nails clenched inboard by deforming the tip over a thick metal rove, in 'Viking' fashion.

Where there is a rapid change of curvature in the hull sections (as, for example, around the bilge), the edges of adjacent planks have to be bevelled to make a good fit. By lashing two small pieces of bamboo together so that they are a bevel breadth apart, and dipping the lower one in ink, the boatbuilder can transfer the shape and the angle needed onto the plank to be fitted, and the bevel is then cut with a sharp and very wide chisel. The bevel required on the outer faces of frames towards the ends of the boat is determined by eye. When necessary, planks are bent by treating them with a fish oil and suspending them, with weights at one end, over a slow-burning fire. Levers with grommets are used to force planks to conform to the curves of the framework until they can be fastened in place.

As planking-up reaches the upper bow, a number of stealers (additional planks) are introduced to give the boat a rising sheerline (see Fig. 7.12). Occasionally, stealers are also used at the stern. It is difficult to maintain plank symmetry, port and starboard, in such circumstances and indeed the planking pattern on the starboard side of a *vattai* seldom matches that on the port side, although hull symmetry is maintained. This additional planking gives the present-day *vattai* a bow which rises much more steeply than that of the *vallam*, and also much more than that of the *vattai* recorded by Hornell in 1914 (see Figs 7.3, 7.4). It is not known whether this change in sheerline is simply to keep the boat dryer in adverse conditions or whether there is a more fundamental reason. Some informants suggested it might be the result of changes in fashion.

When the hull is fully planked the outer faces are adzed to remove high spots and leave a smooth hull.

The crossbeams

With the planking-up complete, the heads of the futtocks, from the sixth to the twenty-fourth, are trimmed to give a level sheerline and a beam shelf, about 100 mm wide and 50 mm deep, is laid on top of them. The upper face of this beam shelf is at the same level as the top strake (yet to be fitted). On top of this shelf are laid eleven crossbeams of varying thicknesses and breadths, from 33 by 180 mm to 48 by 236 mm, but reduced slightly in thickness on their undersides where they lie on the shelf. They are nailed to the futtock heads through the beam shelf (which is the only way in which the latter is secured). All but the tenth and eleventh (counting from the bow) crossbeams extend outboard of the boat. These last two beams act as supports for a small stern deck composed of five or six planks laid on, and nailed to, the beams. Similarly, the sixth and seventh crossbeams support a small midships deck. These decks may not be fitted until all other construction and caulking is complete.

The third, seventh and ninth crossbeams have holes some 83 mm in diameter, worked either centrally or on the after edge, to take the three masts (Fig. 7.12). The floor timbers immediately below these crossbeams (or other timbers fastened adjacent to them) have smaller, shallow holes in which the masts are stepped. Although not observed, we believe that these holes are worked *in situ* and not before installation.

Above the tenth crossbeam, and above the decking that lies upon it, a curved timber (a natural crook), circular in cross-section with a diameter of around 100 mm, is positioned (see Fig. 7.23). This is a timber to which running and standing rigging can be made fast. It has a general, but not specific, resemblance to a 'horse': a raised metal rod, fitted across some European vessels, and along which the sheets of a sail can slide by means of a traveller.

The top strake

Keeping the 'horse' and crossbeams in position is the top strake. This is a narrow plank about 100 mm thick, notched to lie over the protruding crossbeams. As, for most of the length of the boat, this top strake is above the futtock heads, there is no surface to which to nail it: it is therefore sewn to the next highest strake. Pairs of holes are bored through the top strake and the strake immediately below (and through the beam shelf), between 60 and 80 mm apart and with a spacing of anything between 260 and 420 mm. A couple of turns of polypropylene rope are made with a running line, binding each pair of holes and thus fastening this strake to the main hull planking.

The balance board and other equipment

When the top strake has been fastened, two large cradles, 50 mm in thickness and around 700 mm in length, are nailed outboard of the thirteenth frame (see Figs 7.12, 7.13). These are notched at both ends to fit the outboard projections of the seventh crossbeam. Their top surface is also notched and angled upwards towards the bow, at approximately 10 degrees, to take the balance board, the boat's most distinctive fitting. Balance boards are made of *vakai* and are of a length greater than three times the beam of the boat (Fig. 7.18). One example measured 6.48 m in length, 0.305 m in breadth and was 40 mm thick; others were said to be 7.32 m by 0.305 m by 51 mm. Occasionally balance boards are composite, with extra planking added at the ends (Fig. 7.19).

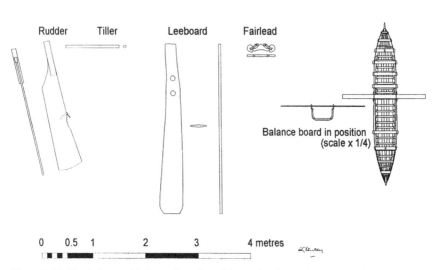

Balance board

Rudder Tiller Leeboard Fairlead

Balance board in position
(scale x 1/4)

0 0.5 1 2 3 4 metres

Figure 7.18 Typical *vattai* balance board, rudder and tiller, leeboard and fairlead.

Figure 7.19 Demonstration of the rowing position in a *vattai*. The starboard end of a composite balance board can be seen in the foreground. In the bows the anchor rope has been led outboard through a fairlead.

Also shown in Figure 7.18 are a rudder and tiller: as Hornell (1945: 224) observed 80 years ago, the *vattai*'s rudder is hung from a single pintle but also lashed through a hole in the stern-post head. A typical leeboard is also illustrated in that figure: *vattai* usually carry three of these.

Caulking and paying

The construction is now essentially complete. The final process is caulking: coconut fibre (coir) is forced into the seams between the planks and the seams are then coated with a hot, tar-like substance, *kunkilia*. This is done in the boatbuilder's yard by a specialist caulker. At Sharif's Atirampattinam yard seams were payed (i.e. waterproofed – see Glossary) with a red aromatic paste, a combination of heated groundnut oil with an unidentified incense-like substance. The upper hull is then payed with a fish oil and the underwater parts of the hull with tar.

Vattai *variants*

In a boatyard at Nambu Thalai, just north of Thondi, a *vattai* under construction in 1999 (alongside a motorised *vallam*) was being prepared for an engine. A second motorised *vattai* was said to be at sea. We were not able to ascertain the reasoning behind this modification, nor could we evaluate how successful this adaptation had been. No other motorised *vattai* was observed or mentioned by informants.

The typical *vattai* recorded in 1997 had a horizontal central bottom plank (foundation plank), and thus a longitudinally flat bottom from the stern-post scarf to the scarf with the lower stem. However, on boats being built in 1999 in Sharif's yard at Atirampattinam (not visited in 1994 or 1997) the forward end of the foundation plank and the intermediate timber between the foundation plank and the stern post were both curved upwards. Aft, this gave a rise of 4 inches (100 mm) at a point 62 inches (1.57 m) forward of the stern post; while forward, there was a corresponding rise of 6 inches (150 mm). Such a modification would make the boat easier to move on the foreshore and more responsive to the helm when underway. It is uncertain whether this was, in fact, the reason for this change. It may be that it was believed to 'look good', which seems to be the reason why many recently built *vattai* have an exaggeratedly high bow.

Ritual and decoration

After the foundation plank of a *vattai* is laid, the boatbuilding procedures and the boat herself are ceremonially blessed. The *mistri* first pours a bowl of saffron water over both ends of the plank. He then takes a hand axe and makes three cuts in the plank, removing a small piece of wood which he gives to the owner of the boat. The *mistri* of a *thoni* at Tuticorin performs a comparable ritual by placing turmeric on the keel and then breaking coconuts over it. On *vattai* being built at Atirampattinam, faces were subsequently drawn in white chalk on the planking and stem post to ward off evil.

When a *vattai* has been completed, a garland of flowers is placed over the stem post. The *mistri* then takes a small bowl of saffron water (or sometimes milk), adds three *betel* leaves and sprinkles the mixture over the stem and stern posts. A lemon is then cut in half and red powder, *kumkum* (with which Hindus mark their forehead), is sprinkled on the two halves. The *mistri* then squeezes the lemon over the stem post, giving the impression of anointing the stem with blood. The owner subsequently gives the *mistri* a *lungi* (loin cloth) and, if he is rich, *datchanai*, i.e. a gratuity.

A similar ritual is undertaken before launching a Tuticorin *thoni*. The *mistri* pours a bowl of milk and seawater over the bow of the ship while a Christian prayer is said; and the owner presents gifts to the workmen, the *mistri* and the *tindal* (the ship's master).

The planks of a *vattai* above the waterline are commonly painted in bright, primary colours. A different, more striking, colour (often red) is normally applied to the posts, and spiral and floral motifs are painted on the sheer strakes. Christians may paint the sign of the cross on the stem post; while just aft of this post an eye (perhaps the 'evil' eye, or oculus) may be depicted. Hindu, Muslim and Christian motifs may be seen on boats which are lying alongside one another. The stem heads of some *vattai* were carved into a distinctive finial: we were told that one type was to be found on Muslim boats and another on those owned by Christians (Fig. 7.20). The owner's name, and sometimes that of the artist with his qualifications, is frequently painted on the bow planking.

Other motifs seen on the upper planking, usually towards the bow, are birds and religious buildings. On the midships planking, where the balance board is fitted, a brightly coloured diamond shape is often painted. Hornell

Muslim Boats   **Christian Boats**

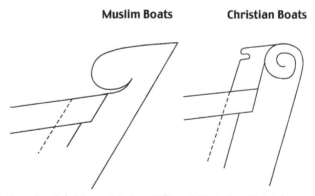

Figure 7.20 Stem-head finials on Muslim (left) and Christian (right) boats.

(1945: 224) refers to ornaments of 'crude and conventional devices' that he saw in the early twentieth century on boats at Muthepet and adjacent villages: '... the most frequent are diamonds, flags, flower sprays and birds in flight ... some have a geometrical figure on the bows and quarters suggestive of a conventional relic of what were originally "eyes"'.

Boat operations

The crew

In the early twentieth century, Hornell (1945: 223) noted that the *vala vattai* of Muthepet had a five-man crew. The basic crew of a *vattai* today can consist of just three men: one to steer and two to handle the sails. In the boat in which we sailed, an older man was steersman and evidently in command (see also Rajamanickam and Jayakumar, 1991: 49); the other two worked the rigging and, when necessary, stood or squatted on the balance board. Extra men are needed when a boat is launched and when going fishing.

Under oars

One or two oars – long spars with a short, narrow and rounded piece of wood lashed to one end as a blade – can be manned on each side of the boat. These oars are used through a rope grommet led through a hole in a kevel head (a framing timber extending above the sheerline). Each oarsman sits on a crossbeam/thwart facing aft, braces his feet against the next beam astern and pulls one oar (see Fig. 7.19). Oars are sometimes stuck into the foreshore, blade uppermost, apparently to mark a boat's particular berth.

Hornell (1910: 46) stated that Muthepet boats were 'paddled', but there is no sign of paddles today, and it seems probable that he meant that oars were used without over-exertion.

Under sail

The small *vattai* have one mast stepped amidships (Fig. 7.21); the larger ones have three masts, each rigged with a single lateen (Fig. 7.22; see also Fig. 7.9). The full rig was seen only occasionally, and *vattai* were normally propelled by sails on the fore and main masts only. As is common in the autumn and winter along the whole of this eastern coast of India, Tamil fishermen tend to sail to their fishing grounds with the offshore land breeze in the early morning and return with the onshore sea breeze in the early afternoon. In the normal course of events, therefore, the *vattai* employed on fishing sails with the wind in the stern sector, both out and return, and seldom has to steer a weatherly course. Under well-set fore and main with a quartering breeze a *vattai* makes good progress. In light winds, and generally when trawling,

Figure 7.21 A small *vattai* under sail off Thondi.

three sails are set; as the wind increases, first the mizzen and then the foresail are taken in. When altering course, the *vattai* is not tacked through the wind, but, like most other lateen-rigged vessels, is worn round by passing the stern through the wind. During this manoeuvre, sails are shifted from old to new lee side by passing the yards forward of their respective masts.

Masts

The three masts are all of teak (*Tectona grandis*), the foremast being about nine-tenths the height of the main, the mizzen about six-tenths (Fig. 7.22).

220

Figure 7.22 A *vattai* with three masts beached at Eripurakarai.

The heel of each mast is stepped in a hole worked in a floor or in a transverse or a fore-and-aft timber set immediately aft of a floor: the fourth, thirteenth and twenty-second from forward. The main mast has lateral support from wooden mast partners either side of its step. All three masts are supported higher up by a crossbeam: the third, the seventh and the ninth from forward (see Fig. 7.12). The mizzen is inserted through a circular hole in its beam, whereas the mainmast and foremast are set within semicircular recesses cut out of the after edge of their beams. Each mast is held to its mast beam by a rope lashing which is tightened by a Spanish windlass with its bar lashed to the mast or to the balance board.

Fastening points

Some of the running and standing rigging (now made from synthetic materials, but formerly of coconut fibre or hemp – see Rajamanickam and Jayakumar, 1991) is made fast to crossbeams: on or near the centreline for standing rigging; on any convenient part, including that protruding outboard, for the running rigging. Some of the beams have a vertical groove worked in their after and forward edges outboard, turning them into cleats to which lines can be fastened. A naturally curved timber (the 'horse') is set

221

across the boat above the tenth beam, and rigging is also fastened there (Fig. 7.23). Some *vattai* have a second, smaller 'horse' across the heads of the aftermost frame. Shrouds are taken to the outer parts of the weather balance board, thereby giving them an optimum angle with the mast and a broad shroud base.

Standing rigging

The three masts are rigged as follows (see Fig. 7.22):

Foremast: A forestay to the first or second crossbeam and a backstay to the 'horse'. An angled shroud to the windward balance board.

Mainmast: A forestay to the third crossbeam and one, sometimes two, backstays to the 'horse'. A shroud to the balance board.

Mizzen: No forestay is fitted. The backstay is taken to the post or to the after 'horse'. Angled shroud to balance board.

Figure 7.23 At sea in Palk Bay. The *vattai* steersman has his right hand on the tiller and his left foot on the curved 'horse' timber to which rigging lines are fastened.

Sail and yard

The three sails are high-peaked lateens of settee shape (see Glossary) with a short luff (forward edge) (see Fig. 7.9). They are made of cotton and are generally a russet-brown colour due to the preservation process when the sail is boiled in a mixture of flour, tamarind-nut husks and the bark of several trees (Rajamanickam and Jayakumar, 1991: 74–5). A bolt rope is sewn to all edges of the sail, and the head is laced to a long, curving bamboo yard held to the mast by a traveller. The halyard, by which yard and sail are hoisted and lowered, is made fast to a cleat at head-height on the main and fore masts; the mizzen halyard appears to be taken to the main 'horse' or to the tenth beam.

Running rigging

The sails are named, from forward, foresail/*aniyattupai*, mainsail/*nadupai* and mizzen/*pichchilpai*. Each one has a tack from its forward lower corner and a sheet from its after lower corner. Precisely where these lines are made fast depends on the point of sail, which determines the angle of each sail relative to the boat.

Foresail: Tack to the bows; sheet to the mainmast beam.

Mainmast: Tack to fifth or sixth beam; sheet to 'horse'.

Mizzen: Tack to eighth or ninth beam; sheet to post or to after 'horse'. Occasionally, a bowline from mainsail to foremast has been observed.

Use of the balance board (see Figs 7.18, 7.19)

The balance board (*kadisu* or *vari*), lashed in its two cradles, generally has two sets of ballast on its starboard arm: a group of stones hanging from the board close to the boat's side; and a 'sandbag' lashed to the upper surface of the board, well outboard (Fig. 7.24). This apparently permanent configuration suggests that, in the winter season, the *vattai* is mainly sailed with the wind from starboard. This would seem to be a practicable arrangement in her main role when she sails a steady course with a generally fair wind, both to and from the fishing grounds, wearing round rather than tacking when course has to be altered.

To further counteract the heeling forces imposed on the boat due to the wind in the sails, one or two of the crew (as required by the prevailing conditions) stand or crouch on the weather balance board, taking a position which keeps the boat as near upright as possible. The crew also find the

Figure 7.24 At sea in Palk Bay with one of the crew of the three-man *vattai* standing on the starboard side of the balance board. A plastic bag containing ballast is lashed to the outermost part of the board.

balance board a convenient place to sit and wash the mud from their legs and feet after they have pushed the boat across the foreshore at launch.

The balance boards of the larger boats are set on angled cradles (Fig. 7.12), which not only raise them above the water but also tilt their leading edge upwards, thereby reducing the chance of the board striking the water and destabilising the boat. Should a board touch the water, it tends to plane rather than dig in, or at least generates a lift force which brings the board to the surface (Hasler, 1950).

Instead of the plank-like balance boards of the larger boats, many of the smaller *vattai* have a simple spar (Fig. 7.21) which extends outboard on each side by approximately a boat's breadth; rigging is made fast to such spars. Since these smaller boats are sailed by just one man, it is impracticable to stand on these spars to balance the boat. However, ballast is loaded on them, as in the larger boats, and moved from the old to the new windward side on completion of wearing.

Changes in the sailing rig

The rig in 1914 of the three-masted Muthupet and Atirampattinam *vala vattai* (Figs 7.3, 7.4) published by Hornell (1945: figs 4 and 6) is not precisely the same as the rig of today's *vattai* (Figs 7.9, 7.22). The masts are stepped in the same relative positions (16, 46 and 77 per cent of the overall length from the bow), but they are of different relative lengths. Whereas the foremast of the 1914 *vala vattai* was 61 to 64 per cent the length of the mainmast, and the mizzen 76 to 78 per cent, the foremast of the 1997 *vattai* is 88 per cent and the mizzen 64 per cent. In other words, during the intervening 80 years, the foremast has become relatively longer than the mizzen mast. This difference is reflected in the relative size, and hence importance, of the foresail and mizzen. The foresail is now only slightly smaller in area than the mainsail (Fig. 7.9), whereas in 1914 the foresail was considerably less and, in fact, smaller than the mizzen (Figs 7.3, 7.4).

The shape of the sails has also changed during this period. In 1914 the *vala vattai* had quadrilateral sails (Figs 7.3, 7.4), similar in shape to those of the late-twentieth-century Orissan *patia* (Blue *et al.*, 1997), which is probably best described as a form of lug, although it is handled more like a lateen than a lugsail. Today the *vattai* also has a quadrilateral sail, but it is set on a longer and more flexible yard (Fig. 7.9). Furthermore, its leading edge (luff) is much shorter than that of the lugsail, and the after edge (leach) rises to a much higher peak; such a sail is best described as a settee lateen.

In general, the centre of effort of a settee lateen is further aft than that of a lug sail. The shift from a lug-shaped sail to a lateen could therefore have made it possible to achieve sail balance with a smaller mizzen sail or, in certain conditions, without a mizzen at all, thus reducing the crew's workload. The change to a lateen rig may also indicate an intention to improve performance in light airs and possibly sail somewhat closer to the wind.

Leeboards

Since such a flat-bottomed hull can have little resistance to leeway, the *vattai* carries two or even three leeboards (*kadappalagai*) – blade-shaped wooden boards (see Fig. 7.18) over 3 m in length. These are slung over the lee side, near amidships, to counteract drift to leeward, i.e. downwind. Rajamanickam and Jayakumar (1991: 63) reported that they were also used to alter course by varying the number used and their position. They could also be used to achieve sail balance.

Steering arrangements

Hornell (1945: 224) noted that the rudder of the Muthupet *vala vattai* was 'large and powerful, hung by gudgeon and pintle at the lower end, and by a coir lashing below the tiller' (Fig. 7.3). The rudders of all *vattai* seen in 1997 were of this type (though not so ornate – but see Fig. 7.25), shaped to match a straighter stern post (Fig. 7.12) and sometimes with a skeg at the foot extending aft. The tiller, which is *c.* 1 m in length, passes to starboard of the sternpost, and the helmsman sits on or leans against the port planking with his feet on the 'horse' where much of the rigging is to hand (Fig. 7.23).

At Atirampattinam, unlike elsewhere, it was noted that, on most (possibly all) *vattai*, the rudder pintle had been bent by *c.* 20 degrees from the rudder's axis of rotation: this had the effect of rotating the axis of the pintle forward and outboard. This had been done so that the rudder could be used in shoal water at that particular site. Several boats were seen under sail negotiating the narrow, shallow channels which are characteristic of the Atirampattinam landing place. The rudders of these boats had been shipped on their bent pintle, but without the upper lashing. They were then used as a side rudder, sometimes with a tiller shipped aft, by bringing the rudder head forward to the quarter. The resultant tilt of the rudder reduces its draft and inclines the leading edge, thus allowing it to trail through the muddy seabed.

When sailing stern first from the foreshore, a leeboard is slung over the (old) port bow and used to steer the boat from the (new) starboard quarter, with a short spar inserted into a hole near its upper edge as a tiller.

Use of landing places

On the northern shores of Palk Bay, *vattai* are used from beach landing places which, at low water, are often a considerable distance from the sea. Even at high water there remains a broad foreshore of thick estuarine mud (Hornell, 1945: 227). Depending on the state of the tide and the boat's draft, *vattai* returning to the landing place at Eripurakarai can usually be sailed to, or close to, their particular berth, to take the ground with bows pointing inland (see Fig. 7.9).

When leaving these berths, *vattai* almost invariably have to be manhandled to the sea. With an offshore wind, as is usually the case, sails are set to propel the boat astern and the crew, with helpers, push on the balance boards and on the projecting beam ends. Once afloat, the boat continues to be sailed stern first until well clear of the beach, with the helmsman using a leeboard as a rudder. Poles are used to propel *vattai* where the seabed is firm, but the Eripurakarai foreshore is too mobile for them to be used.

In 1997 the eastern stretch of the landing place at Eripurakarai had relatively firm sand around and above high water mark. In 1999, this had been overlaid or replaced by soft mobile sediments and, as a result, few *vattai* were berthed there. The majority of the *vattai* were beached to the west of this, further away from the access road but where the beach above spring tides remains firm. For all these boats, however, there still remains a stretch of soft foreshore over which they are man-hauled.

When questioned about the difficulties of such operations, Eripurakarai fishermen indicated that they could cope and that if things got worse they would adapt; they had no knowledge or folk memory of any time when it was easier to use the Eripurakarai landing place. Their reaction to the suggestion that beach capstans might improve their lot was that these capstans were only used at Mallipattnam. Since that landing place is further away from the Kaveri (Cauvery) delta, it has, in fact, a much less mobile foreshore than Eripurakarai.

Hornell (1945: 222–4) had noted that Muthupet, some 20 km to the northeast of Atirampattinam, was a landing place in the early twentieth century. There are still a few *vattai* working out of Muthupet, and at least one boatbuilder. These particular *vattai* still retain the ornate rudder (Fig. 7.25) noted by Hornell in 1914 (1945: fig. 4) (see Fig. 7.3).

Vattai berths at Atirampattinam are even further from the sea than at Eripurakarai and boats there each have an individual channel across the mud to their berth. As Hornell (1945: 227) described them, 'at low water these long rutted channels, 2–3ft. wide, make the uncovered mud-flats look like a great railway yard with many sidings'.

When berthed on the mud, boats may be made fast to a spar stuck into the foreshore. An anchor, of a type universally known as a 'fisherman's anchor', is sometimes let go some distance to seaward of the boat's berth so that it can be used later to haul the boat stern first off the beach. Anchors are also used in the more conventional way, from the bows. An ornate piece of wood (see Figs 7.18, 7.19) with a curved upper edge is lashed across the bows above the foremost frame and this acts as a fairlead for the anchor line, the inboard end of which is taken to the second crossbeam.

Figure 7.25 A *vattai* at Muthupet in 1999, with an ornamental rudder similar to that depicted by Hornell in 1914 (see Fig. 7.3).

Fishing

Gill nets and trawl nets were both used by *vattai* fishermen at Eripurakarai; frequently, both types were used from one boat. Line fishing gear was also seen in some boats. The fine-mesh gill net, with sinkers and floats, is set from three Dan buoys in about 4 m of water, the depth having been established by sounding pole. If the *vattai* is also trawling, the gill net is left unattended until later in the day.

Trawl nets, used mainly to catch prawns, are set to windward from spars projecting forward and astern of the boat. Under the appropriate sail (usually all three), and with ballast on the windward balance board, the *vattai* is then sailed in a controlled drift to leeward.

During the winter, generally on this coast, fishing boats regularly sail at about 05.30 hours and return around 15.00 hours. This is probably so that they can use the land breeze in the morning (in January this is from the northwest to the north) and the sea breeze in the afternoon (from a more easterly direction).

The tidal range is so small (the mean in Palk Bay is *c.* 0.76 m) that, even with the very slight beach declivity that there is at Eripurakarai, for example, the point at which a boat will float, and from and to which it has to be dragged, does not vary much between high and low water, even at springs. Furthermore, the tidal flows are weak. Thus varying fishing hours to match the ebb and flow of the tide would gain little, if any, advantage in terms of the distance boats had to be manhandled or in terms of fair tidal flows.

The question of the times of launch and landing needs further evaluation at other landing places on this coast and at other seasons.

Discussion

A Tamil tradition

The three types of boat (*vattai*, *vallam* and *vattal*) and the two ships (Tutticorin *thoni* and Cuddalore *kotia*) of Tamil Nadu considered in this paper differ in size and function. They have, however, other characteristics which are so similar that, at a certain level of classification, these boats and ships may be considered variant forms of a single class of planked vessel. These common characteristics may be discussed under four headings: shape; hull structure; means of propulsion; and design.

Shape

All are (near) double-ended in plan and have a box-like transverse section, being flat (or flattish) in the floors and (near) wall-sided over a large proportion of their length. This is a form which gives maximum space for cargo etc.

Structure

All are built 'frame first', i.e. elements of the framing define the shape of the hull. This framework is first erected, then the planking is fastened to it. The framework consists of four elements:

i)   A master frame, placed at/near the midships station, that is, at/near the mid-point of the keel/centre bottom plank.

ii) A number of 'equal' frames which are identically equal to the master frame and which are fitted forward and aft of it.

iii) A number of 'unequal' frames, each pair of which differs from the master frame by a designed amount. These are fitted beyond the equal frames, one of each pair forward of midships, one aft. The foremost and aftermost of these unequal frames have a special name: in the case of the *thoni*, this is *rodhai*; in the *vattai*, it is *kutchaicy*.

iv) A few passive frames at the ends of the vessel (not in the *vallam*).

The master frame, the other equal frames and the unequal frames are all active: that is, they are designed and they define the vessel's shape. In general terms, the number of equal frames determines the length of the main hull and the carrying capacity of the vessel. The number of unequal frames and their individual form determine, with the posts, the three-dimensional shape of her bow and stern. The passive frames at bow and stern are not designed: their shape is obtained by spiling – a term which, in this context, means transferring the hull shape onto patterns so that framing timbers can be fashioned to match the hull – from post, plank and keel or centre bottom plank, after some of the planking has been fastened to the framework.

In the *vattai* the passive frames are half-frames which meet and/or slightly overlap on the middle line. The active frames in the boats each consist of a floor timber and a pair of futtocks fastened together, before installation, in a complex joint which includes a dovetail element (see Fig. 7.17). The ships' active frames have top timbers, in addition to futtocks, and these five timbers are joined by similar complex joints.

In the ships, the planking is fastened to the framing by bolts with nuts on the inboard faces of the frames. Bolts may also be used in the boats, but generally the boat planking is fastened to the framing by nails driven from outboard and clenched inboard by turning their emerging tips back into the inner face of the frame. The passive frames are fastened to the planking in a similar manner.

The plank-keel or centre bottom plank is joined to the posts in a simple horizontal box scarf and fastened by nails hooked as in the planking. Planks within strakes meet in butt joints at frames.

Away from the main hold, crossbeams are supported by stringers fastened to the inner faces of frames (ships) or by beamshelves at the frame heads (boats). In the boats most of these crossbeams can be used as thwarts.

Propulsion

Apart from the *vattal*, which is a dumb (towed) lighter, these ships and boats are propelled by settee-lateen sails set on one or more masts. In addition, the *vallam* and the *vattai* have an outfit of oars.

Design methods

It is clear that the method of designing the two ship types is generally similar to that used by boatbuilders. The design, building and use of the Atirampattinam *vattai* was the main focus of the fieldwork in Tamil Nadu: this will therefore be the design method generally discussed here, with occasional remarks about the other vessels. On present evidence, this design method is not found elsewhere in the subcontinent.

The fundamental unit is the length of the plank-keel or centre bottom plank, which is itself related in a known way to the overall size/cargo capacity of the vessel to be built. Ratios based on this length give the maximum beam and depth of the hull at the station of the master frame. The rake of the *thoni* posts is known in terms of a table of co-ordinates. Since the main frames are spaced at a standard distance apart, the number of frames along the length of the keel can be calculated.

The shape of the *vattai*'s master frame is given by a full-size wooden mould of somewhat more than one half of the frame, extending beyond the centreline. This mould is evidently a key 'mystery' in the builder's art. The shape of the equal and the unequal frames are derived from this mould using a scrieve board or a mould loft floor. The main parameters of such an empirically derived, three-dimensional hull shape are encapsulated within the *vattai* design system in two ways:

(a) By ratios of principal dimensions: this ensures that all vessels built will have similar midships sections and be the same general shape – a shape that has proved successful in the past.
(b) By simple geometric means the two-dimensional shapes of each pair of unequal frames is derived from the master frame, so that the three-dimensional narrowing and rising of the hull is controlled, thereby giving specific shapes to the ends of each vessel. The proportion of hull length with designed frames varies from 100 per cent in the *vallam* to *c.* 60 per cent in the *thoni*.

By varying the keel length and the number of equal frames, the designed cargo capacity can be altered. And by varying the number of unequal frames, the shape of bow and stern, of entry and run, can be varied, with consequences for the vessel's performance. With a fixed keel length it is possible to produce different shapes of hull by varying the ratio of the numbers of equal to

230

unequal frames. Theoretically it would be possible to design a non-double-ended hull by having a different number of frames aft from forward. In other words, the master frame would no longer be amidships (cf. the description of a thirteenth-century Brindisi galley noted by Bellabarba [1996: 259]); a vessel designed in this manner could, for example, be fuller aft than forward. It is not known whether such a design modification has ever been used in Tamil Nadu.

Comparison with other accounts of design and building frame first

Mediterranean moulding techniques

The earliest documentary references to methods of designing the framework of vessels built frame first comes from fifteenth-century Venice where the *partison* or 'rules' method was used (Bellabarba, 1993; Johnston, 1994); there are indications, however, that this design system was used as early as the thirteenth century, and even earlier (Bellabarba, 1996; Rieth, 1996). These design methods have come to be known as 'Mediterranean Moulding' (Sarsfield, 1988: 71).

By the sixteenth century comparable design methods were used in Spain, Portugal, France, southern Netherlands and England (Bellabarba, 1993: 286, 288, 290); the variant methods used in these countries are now known collectively as 'Atlantic'. Mediterranean methods, or ones derived from them, were being used in the Ukraine, northern Italy and Provence in the mid-nineteenth century (Bellabarba, 1993: 286). In the late twentieth century, European-derived methods were documented in Newfoundland, Brazil and Greece (Sarsfield, 1985: 88; Taylor, 1988; Rieth, 1996: 177–99).

In this method of designing a vessel the main dimensions of the hull were known as proportions of a modular unit, usually the keel length or the length overall. Bellabarba (1993: 274) does not give examples of these ratios, but Grenier (1994: 137–8) has noted the sixteenth-century Basque use of a rule giving the ideal ratios as:

Bmax:Lk:Loa = 1:2:3

where: Bmax = maximum beam
        Lk = length of keel
        Loa = length overall

In the Mediterranean region generally, the designed part of the hull was usually somewhat less than the keel length; the last designed frames had a special name (in Italian, *capo di sesto*). Beyond this point the hull shape evidently changed too rapidly for shapes to be calculated by the *partison* methods (Bellabarba, 1993: 282). The bow and stern frames were thus passive, their shape being determined late in the sequence, probably by battens between the two posts.

The total number of designed frames and the total rising and narrowing of the hull between midships and the two *capo di sesto* were first calculated. The shape of the master frame (near amidships) was encapsulated in a 'rule' which gave the orthogonal co-ordinates of its curve at four points (Bellabarba, 1993: fig. 3). This shape was evidently marked out full size on a wooden pattern/template or mould by connecting these points with a flexible batten. The master frame was fashioned from this mould and other frames (equal frames) were then made to the same shape, the number depending on the length of dead flat hull required.

The shape of the remaining designed frames (i.e. the unequal frames) was obtained by simple geometric means from the master mould. A series of horizontal lines, equal to half the number of unequal frames, was inscribed on a semicircular wooden tablet known as *mezza luna* (half moon) in Italian, *tablette* in French, using a simple geometric construction which made the intervals between these lines decrease steadily in accordance with a geometric progression (Fig. 7.26). These intervals were transferred to a measuring stick (*brusca*), which was then used to mark on the mould of the master floor timber the appropriate rising (*stella*) and narrowing (*de fondi*) and thus give the floor shape of each pair of unequal frames in turn. The half moon was similarly used to mark out the appropriate curve at the bilge where floor joined futtock. It could also be used to allow for a widening of the futtocks (*del ramo*) at a higher level (Bellabarba, 1993: 280–81, fig. 4).

With the equal and unequal frames fashioned and fastened to the keel, the shape of the main hull was established and planking-up began.

Design in twentieth-century Brazil

Sarsfield (1985, 1988) has documented similar design methods recently used to build *escunas* or *saveiros*, small sailing vessels 10 to 30 m in length, in Bahia, Brazil. Since these 'schooners' have a transom stern and a more curvaceous shape than the Tamil Nadu *vattai*, the design system allows more refinements to shape.

The general concept of the required hull shape is encapsulated in ratios relating the length between posts at deck level (L) to the height (Ht) and breadth (B) of the midship or master frame. The total rising and narrowing at bow and stern are also known in terms of (L). Typical ratios are:

$B = L/3$
Ht approx. $= 2B/5$
Floor of the midship frame $= B/2$
Total rising or narrowing of all the unequal frames $= G = L/72 = B/24$

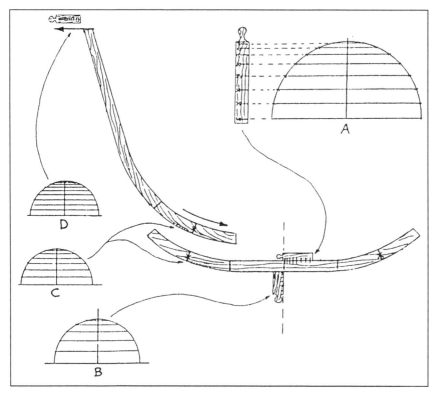

Figure 7.26 Diagram illustrating the *partison* or rules used with a wooden tablet, known as the *mezza luna* (half moon), to design the frames of a great galley in fifteenth-century Venice. After Bellabarba, 1993: fig. 4. A. The narrowing (*fondi*) of the floor timbers. B. The rising (*stella*) of the floor timbers. C. Fairing the junction of futtock and floor (*scorer del sesto*). D. Widening of the futtocks (*ramo*).

Sarsfield does not note how the keel length is related to these dimensions, but it may have been 2L/3, as in the sixteenth-century Basque region (Grenier *et al.*, 1994: 137–8).

These proportions are used to draw at full size the shape of the master frame on the mould loft floor, within a rectangle of sides (L) and (B). The curve between the end of the floor and the upper corner of this rectangle is drawn by eye using a flexible batten (Sarsfield, 1988: 65, fig. 1). Two moulds, one for the floor and one for the futtock, which overlap at the bilge, are made from this drawing. These moulds are for slightly more than one side of the master frame and extend beyond the centreline, as on the *thoni*. A master frame and two or three equal frames are made from these moulds.

The group of unequal frames extends forward and aft only as far as a point half-way between the master frame and the ends of the keel. The foremost and aftermost of these designed frames have a specific name, *almogamas* (Sarsfield, 1988: 66). The shapes of all these frames are individually derived from the master moulds using a diagram similar to the Mediterranean *mezza luna*. As in the Mediterranean and generally in Tamil Nadu, the bow and stern frames are not designed but are passive: their shape is derived by spiling from a number of battens which run from bow to stern and are adjusted by eye.

The Brazilian builders use the *graminho* (the *mezza luna* scale inscribed on the tablet or tabua board) not only to design the unequal frames (Fig. 7.27) but also to mark bevels on the outer faces of the futtocks before these frames are installed, and to mark another bevel on the outer faces of floors of those frames that are to be canted (Sarsfield, 1988: 67–8). These refinements ensure that the minimum of trimming is required when the whole framework is faired after installation.

Sarsfield mentions that Bahian futtocks overlap forward of the forward floors and aft of the after floors (the opposite sense to that used in Tamil Nadu), but he does not describe how they are joined. Evidence from five sixteenth-century ships, three wrecked in American waters and two in British waters, which may all have been built in the Iberian peninsula, is relevant here. Floors in the central part of these sixteenth-century hulls were joined to futtocks by dovetail-shaped mortices and tenons fastened by treenails and spikes (Redknap, 1984; Grenier, 1988; Keith, 1988; Hutchinson, 1991). These dovetailed joints may be compared in complexity to the corresponding joints in Tamil vessels (see Fig. 7.17). The equivalent joint in the medieval Mediterranean was a form of box scarf known as an *écart à cadeau*: such joints were found, for example, in the fourteenth-century wreck Culip 6 off Catalonia in the west of the Gulf of Lyons (Rieth, 1996: 152, fig. 97).

Tamil Nadu, Brazil and Europe

There are some differences between the three systems of design described above, but these are of a minor nature and the systems have much in common. The essential information required for all three includes: a basic length module; the general shape of the vessel in the form of ratios relative to that length; the shape of the master frame; the number of frames; and the total rising and narrowing of the designed part of the hull. In all three methods emphasis is placed on the positioning of the foremost and aftermost designed frames and these are specifically named. The shape of the extreme bow and stern frames is obtained by spiling from ribbands/battens/planking.

The shape of the master frame in the Mediterranean/Atlantic and in Brazil is obtained by eye using a flexible batten, with greater guidance in the Mediterranean from known orthogonal co-ordinates. Since the Tamil Nadu vessels are all wall-sided around amidships, such refinements are unnecessary

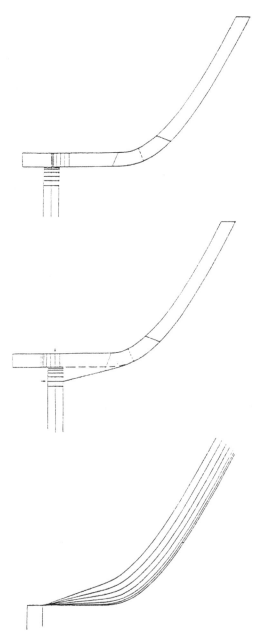

Figure 7.27 Diagram to show the use of a rising board and a composite mould (floor and futtock) in twentieth-century Brazil to obtain the shape of the frames of a small 'schooner'. After Sarsfield, 1988: figs 3, 4 and 5. Top: The shape of the master frame. Middle: The shape of one of the unequal frames. Bottom: The master frame and six unequal frames drawn out.

and the curve at the bilge seems to be derived entirely from handed-on moulds.

The shape of the unequal frames in all three regions is obtained by simple geometric modifications to the shape of the master frame. In the Mediterranean/Atlantic and Brazil, a geometric scale inscribed on the tablet is used in this transformation. The Tamil design being much less curvaceous, memorised or written tables of the necessary rising and narrowing of the hull are used in conjunction with the scrieve board or mould loft floor. There appears to be no evidence in the Mediterranean/Atlantic or in Tamil Nadu for the Brazilian use of a scale or table to mark bevels on futtocks and floors.

In Brazil it appears that individual moulds are made of the frame shapes obtained in this way, and these are used to mark timbers. In the Mediterranean/Atlantic and in Tamil Nadu, a measuring stick is used to position the single mould on each timber so that the appropriate shape can be marked out.

In Tamil Nadu, a specially complex joint (including a dovetail element) is used between floor and futtock in the designed frames. In the Mediterranean this was a box scarf; however, the evidence from several sixteenth-century wrecks suggests that a dovetail was used in Iberian shipyards on the Atlantic coast, at least in the later period. It is not known what joint was used in Brazil before Sarsfield introduced the dovetail (Carrell and Keith, 1992; Barker, 1993).

The differences noted between these three systems are minor and suggest adaptation rather than independent conception. In particular, the Tamil design system has the appearance of being a simplification of the general Mediterranean method to match the relatively simple hull shapes required. The degree of similarity, identical in some instances, between the three systems of design strongly suggests that the later uses grew out of the earliest use.

Within the zone in which Mediterranean/Atlantic design methods are known to have been used, it is difficult to say from precisely where a version of these methods might have been transferred to India. The dovetail joints in the Tamil frames show affinities with Iberia. On the other hand, the use of diagonal datum lines (functionally comparable to ribbands), as in the Tamil scrieve board diagram (see Fig. 7.15), is not mentioned in Iberian (or Venetian) documents of the fifteenth or sixteenth century, but it is known in seventeenth-century France (Rieth, pers. comm.; 1996: figs 25 and 50). It seems possible, therefore, that the Tamil design method was brought to India by the Portuguese in the sixteenth century or by the French in the seventeenth century.

An assessment of the fieldwork

The aim of the fieldwork undertaken in 1997 and 1999 was to document the *vattai* fishing boats of Tamil Nadu and to study how these boats and other frame-first Tamil vessels were designed. To achieve that aim fully, one would have to become fluent in the local language (especially in technical terms) and spend up to a year on fieldwork. In present circumstances such a research programme is not possible. In two seasons we spent thirty days on fieldwork and we had to rely on an interpreter for all interviews. Although five ships and six boats were studied during building or repair, and innumerable boats examined on the beach and in the boatyard, we were unable to see all phases of the building process; in particular, we never observed the design stage and had to rely on *ad hoc* demonstrations. We went to sea in a *vattai*, but this was for not much more than two hours, and the boat was not undertaking her normal work, nor was the weather such that her performance could be fully demonstrated, although launching the boat was probably as difficult as it ever can be.

Nevertheless, we believe that we have documented the *vattai*, as fully as circumstances permitted, to the standard that Ole Crumlin-Pedersen, the Danish maritime archaeologist, has suggested should be aimed at when recording any boat (McGrail, 1982: 73). As a result a scale model of a typical late-twentieth-century *vattai* (Fig. 7.28) has been built by a professional marine modelmaker of Fakenham, Norfolk, using the report published in *South Asian Studies* (Blue *et al.*, 1998) and archival material.

Furthermore, a Tamil boat and ship building tradition has been identified: these vessels, although of various sizes, functions and operating environments, have shape, structure, propulsion and design characteristics in common.

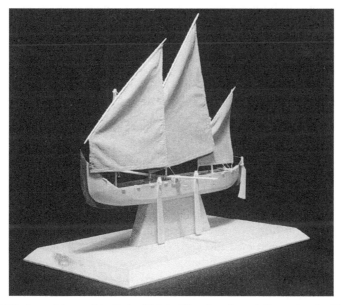

Figure 7.28 A model *vattai* built by Kelvin Thatcher from information in Blue *et al.* (1998), especially figs 10, 11, 15 and 17.

Study of the Tamil methods of designing vessels has led us further afield. These Tamil methods have clear similarities with those recently used in Brazil, but more importantly they can be linked to the frame-first design systems of Mediterranean and Atlantic Europe. The evidence, as it stands now, suggests that the sixteenth-century Portuguese or the seventeenth-century French may well have brought a version of these European practices to Tamil Nadu.

This must remain a hypothesis for the present, since it is not proven that the Tamil building tradition does in fact go back that far. At best, we may say that boats and small ships similar to the *vattai* have probably been designed, built and used in Tamil waters for the past 100 years. In support of that statement we have a not very detailed (and, in one respect, inaccurate) report by James Hornell (1945). There may be earlier accounts and illustrations, yet to be found, that would help to trace this Tamil tradition further back in time, but undoubtedly the most useful evidence would come from some future excavation of an indigenous early frame-first plank boat. Before that can take place the necessary infrastructure needs to be in place.

Tamil maritime ethnoarchaeology and European nautical history

This investigation into the Tamil frame-first building tradition may lead to a greater understanding of late-medieval/post-medieval Europe, when there was a general, though not total, shift from plank-first to frame-first techniques of ship building, with consequent changes from 'by eye' or 'free-arm' techniques to formal design methods.

The differences, significant but not fundamental, noted between the design methods used in the Mediterranean, in Brazil and in Tamil Nadu suggest that the methods documented by Bellabarba (1993) and others (for example Barker, 1988; Rieth, 1996) may not have been the only ones used in medieval and post-medieval Europe. Within general design parameters, variations, both regional and temporal, may well have arisen there to suit local conditions, needs and resources.

The variety of vessels found within the Tamil tradition further emphasises that, within any tradition, there is some variability rather than strict uniformity. Around a core of shared attributes, the Tamil vessels vary in size, function and, to a degree, method of propulsion. Moreover, although latterly *vattai* and *vallam* have been observed on the same foreshore, they are generally used in different environments, leading to two basic subgroups: those that are flat-bottomed and keel-less (the *vattai*); and those that are full-bodied with a plank-keel (the others). There may have been similar disparity within the early European traditions; indeed, the two subgroups found in the Romano-Celtic tradition of NW Europe of the first to fourth centuries AD (McGrail, 1995) are, in this respect, not unlike those of twentieth-century Tamil Nadu.

In contrast with the sewn plank boats of South Asia and other traditional plank-first craft, Tamil frame-first vessels are formally designed, with the production of drawings or plans of the framing. Nevertheless they are not fully designed; there is still an element of building 'by eye'. For example, the passive frames at the extreme ends of the vessels are not designed and the builder has to use his personal knowledge and experience to determine hull shape there. The builder's eye is also used both to fair the main framework before it is planked and to determine the run of the sheerline at bow and stern. The Tamil frame-first builder also uses his 'art and craft' when he works bevels on the unequal frames. This use of personal experience is a major feature, possibly the only 'design' element, in plank-first building. This blend of techniques suggests that there was probably a similar degree of continuity in the late-medieval European ship-building scene, smoothing the transition from 'free arm' plank-first methods, used from time out of mind, to the new techniques needed to design and build a frame-first ship.

There may also be historical lessons to be relearned in the light of the abandonment and re-establishment downstream of Tamil river ports as rivers silted up. From *c.* 3000 BC onwards, European fisherfolk probably faced similar problems to those faced by Tamil fisherfolk on the northern shores of Palk Bay today, as estuaries silted up and bars became established. It may, in future, prove possible to trace the European responses to such environmental changes by studying the abandonment and evident relocation of estuarine and coastal settlements.

Future research

Further fieldwork needs to be undertaken in coastal Tamil Nadu to confirm and, where necessary, amend our findings. This further research, preferably by Tamil scholars, should be undertaken in seasons other than winter and should include detailed work on the maritime environments in which these boats are operated, the sailing and fishing procedures used, and the social and economic milieu within which the catches are marketed.

An ethnographic baseline for the Tamil tradition of frame-first boatbuilding has been established, although certain details of the *thoni* design system need further elucidation. The next stage of research must be to pursue this study back in time by documentary, iconographic and, eventually, archaeological means.

On a wider canvas, it seems possible that the transfer overseas of aspects of late-medieval ship and boat building (in this case, mainly design techniques) by European maritime nations was not limited to Newfoundland, Brazil and India; there may also have been transfers to Africa, Southeast Asia and China. The overseas transmission of European ideas and techniques has been discussed by, for example, Horridge (1979) in relation to Indonesia and Cederlund (1994) for North America, but the process has not yet been

systematically investigated. Should such an international research programme ever be undertaken it would, when published, greatly increase
understanding of how European plank boat and ship building has developed
and how ideas have been transferred to, and adapted in, 'extra-European'
countries during the past 500 years. Such a programme might also identify
ideas which flowed in the reverse direction.

8

HIDE BOATS OF THE RIVER KAVERI, TAMIL NADU

Seán McGrail, Lucy Blue and Colin Palmer

Hide boats are found today in many parts of the world: the Americas; Siberia and Mongolia; Greenland and the Aleutians; Ethiopia; Tibet; India; Arabia; and Britain and Ireland (McGrail, 1998: 173). In earlier times, late Roman authors described their use, both at sea and on rivers, in Arabia, Italy, Spain, Britain and Ireland. Simple forms of this boat type can be made from a single hide: a 'leather bag' reinforced at the rim. Generally, however, a framework of wood, bamboo or, in Siberia, whale bone is first built to the required shape and a 'skin' of several hides sewn together is then fastened to it, often by leather thongs: these boats are thus built frame first. In Britain and Ireland animal hides began to be replaced by tarred canvas from the mid-twentieth century.

Hide boats are, and were, used on rivers and lakes as ferries and for fishing and carrying cargo; at sea, for fishing, inter-island work and hunting seals and walrus. The framework of these boats is shaped to match the role the boat is to undertake and the performance required of it. Thus river boats are often round, elliptical or rectangular; seagoing boats, such as the Irish *curach* and the Inuit *umiak*, are 'boat-shaped'; while specialised Artic boats – the Inuit *kayak* and the Siberian *baidarka* – have an elongated, lanceolate shape which gives them the speed required in their hunting role (McGrail, 1998: 173).

Hide boats in southern India

During our early seasons of fieldwork on India's Bay of Bengal coast, occasional efforts were made to locate examples of hide boats, which Hornell (1933; 1946: 95, 105) and earlier authors noted had been used on rivers in southern India since at least the fourteenth century. These searches were unsuccessful until January 2000 when, during our fifth season, numerous boats were observed on the River Kaveri (Cauvery) at Hogenakal, in northwestern Tamil Nadu, close to the border with Karnataka (Fig. 8.1). This boat landing place (Fig. 8.2) is some 5 km from spectacular falls, on the

241

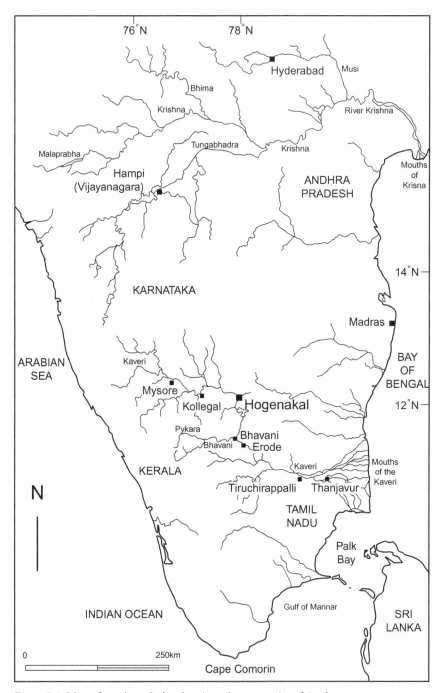

76°N   78°N

Hyderabad   Musi

Bhima

Krishna   River Krishna

Tungabhadra   Krishna

Malaprabha   Mouths
                                                    of
Hampi   ANDHRA   Krisna
(Vijayanagara)   PRADESH

                                                    14°N

KARNATAKA   Madras

ARABIAN   BAY
SEA   Kaveri   OF
                                               BENGAL
            Mysore   12°N
          Kollegal   Hogenakal

          Pykara
                    Bhavani
          Bhavani   Erode

KERALA   Kaveri   Mouths
                                      of the
N   Kaveri
          Tiruchirappalli   Thanjavur

                          TAMIL
                          NADU

                          Palk
                          Bay

          N

                                               SRI
                                               LANKA

INDIAN OCEAN   Gulf of Mannar

0   250km

              Cape Comorin

Figure 8.1 Map of southern India showing places mentioned in the text.

Figure 8.2 Parical near the River Kaveri at Hogenakal.

eastern or left bank of the river, which is fast-flowing, turbulent and fitful, alternating between miserable low water and violent flooding, with granite fragments everywhere obstructing its bed (Deloche, 1994: 35).

In the Tamil language hide boats are called *parical* or *paricu*; in Telugu (spoken in Andhra Pradesh) the term used is *argili* or *putti*; and in Kannada (Karnataka) it is *harigolu* or *butti* (Deloche 1994: 137). These linguistic boundaries, by and large, define the area of use of the hide boat in southern India.

The earliest record of Indian hide boats comes from 1398 when they were used to ferry troops across the River Krisna, or Krishna, in Andhra Pradesh (Deloche, 1994: 138–9). There are three early-sixteenth century inscriptions on stone in Karnataka which refer to *harigolu* used by Anegondi ferrymen to cross the River Tungabhadra at Hampi-Vijayanagara, the capital city of that region, and at seven other nearby sites. In 1520 Domingo Paes described round boats made of cane baketry covered with leather which he had seen ferrying 15 to 20 people across this river at Anegondi (Verghese, 2000: 303–4). In the mid-seventeenth century, Tavernier recorded that large boats of oxhide on a basketry framework were used to ferry goods and people across rivers near Hyderabad-Secunderbad in Andhra Pradesh (Hornell, 1946: 95–6, 105). These craft, which were circular in form and 3 to 6 m in diameter, were propelled by paddles from each quarter of the boat.

In the late eighteenth and in the nineteenth century hide boats were mainly used in the catchment areas of the rivers Krisna and Kaveri (Deloche, 1994:

137–40, figs 1 and 24). At Tiruchirappalli in Tamil Nadu, there were 50 hide boats which, during times of flood, could each carry 4 tonnes of rice down the River Kaveri into the delta. At their destinations they were dismantled, the bamboo framing was sold and the hide retained for further use. This procedure is similar to that used on the River Euphrates in Mesopotamia in the fifth century BC (Herodotus, 1.194) and may be compared with the dismantling of early-twentieth-century buoyed rafts in Baghdad (Hornell, 1946: 28) and on the Yellow River in China (Worcester, 1966: 122).

In the early twentieth century hide boats were still being used on the River Kaveri in Tamil Nadu, in the Hyderabad/Secunderbad region of Andhra Pradesh and on the headwaters of the River Tungabhadra (Fig. 8.3) near Hampi/Vijayanagara in Karnataka (Hornell, 1946: 92–8). At some fishing villages on the Kaveri near Kollegal, Hornell found hide boats which were 'quadrilateral in plan with the corners rounded' and measuring 2.29 x 1.98 x 0.61 m; otherwise all boats that he encountered were circular in plan. In form, they were bowl- or even saucer-shaped and varied in size from one-man fishing boats, 1.5 to 1.8 m in diameter and 0.40 m deep, to boats measuring 4.2 x 1 m which were capable of carrying 50 men or 40 bags of grain. The 'skin' of such boats was made from several oxen or buffalo hides sewn together and then fastened to an open-basketry framework of split bamboo by lashings just below the rim.

Hornell (1933; 1946: 93–6) also noted that, in his day, one-man hide boats were used for fishing on the rivers Pykara, Coleroon and Bhavani on the upper reaches of the Kaveri in Tamil Nadu. Larger ones were used to ferry goods and people across the Kaveri and also across the River Tungabhadra in Karnataka (see Fig. 8.3). Hornell was told by Sir Frederick Nicholson that, in about 1888, he had travelled in a large *parical* on the River Kaveri from Bhavani downstream to Erode and that at that time it was usual for road metal (i.e. stone) and heavy goods to be transported from the interior of Tamil Nadu down the River Kaveri to the plains of Thanjavur in similar boats. These boats were generally paddled, but poles could be used in shallow water. When only a single pole was used the boat was difficult to steer and made erratic progress (Deloche, 1994: 138).

It is not clear how systematic and how widespread were Hornell's surveys of hide boats. He felt able, however, to divide his 'South Indian coracles' into three regional groups (Hornell, 1946: 93–5):

Type A – 'Kaveri'. These were found in the Coimbatore and Thanjavur districts of Tamil Nadu. They were the smallest in size, generally for one-man use, and had only a primary framing.

Type B – 'Tungabhadra' (Fig. 8.3). These were found in Karnataka around Hampi/Vijayanagara. They were larger than the Kaveri type and had secondary, as well as primary, framing.

Figure 8.3 Harigolu at the Anegondi ferry across the River Tungabhadra, Karnataka, in the early twentieth century. From Hornell, 1946: plate 15A.

Type C – 'Krisna'. These were found in the Hyderabad district of Andhra Pradesh. They were 'the largest and finest in India'.

The Hogenakal *parical*

Since no first-hand description of any Indian hide boat had been published since Hornell's in the 1930s, it was decided to seize a fleeting opportunity of about 30 hours between two periods of research on coastal plank boats and travel by car some 200 miles upstream of Thanjavur to Hogenakal, on the River Kaveri, where hide boats known as *parical* were said to be still in use.

The time available on site was only $2\frac{1}{2}$ hours, but by dividing into two teams it proved possible to compile a measured diagram of a representative *parical* (Fig. 8.4), observe a part-built boat being completed, interview boat owner/users and boatbuilders, and make a short sortie on the river in a *parical*. Numerous black-and-white and colour photographs and colour slides were taken.

Our principal informant said that there were ten builders in the vicinity, but this may have included assistants since another informant said that there were only four. Each builder was said to build around 30 boats annually. Boats are built to order, the material being provided by the prospective owner. At any one time during our visit there seemed to be around 70 boats visible in the river and on the shore at Hogenakal; an unknown number were said to be in use elsewhere. One informant claimed that, in his grandfather's time (40–50 years ago?) 100 to 150 fishing boats were used from the village,

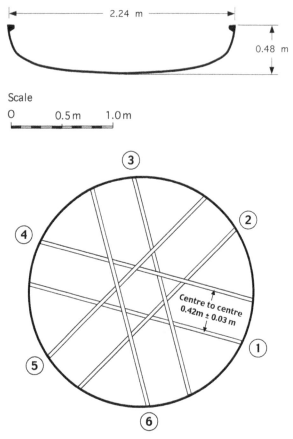

Figure 8.4 Measured cross-section of a typical Hogenakal *parical* and a diagram showing the positioning of the secondary framing.

but nowadays, owing to growing numbers of tourists, there were many more boats, perhaps over 250. The average *parical* was said to last about six months. Assuming that the estimates of boat-life and of boats built per builder were accurate: with only four boatbuilders, a boat population of *c.* 60 would be likely; if ten, a population of *c.* 150 would be possible. In the time available, it was not possible to refine these estimates.

We were told that, with all materials to hand and with a team of three builders, a *parical* could be built in one day between 07.00 and 15.00 hours. This estimate seems reasonable from our observation of the speed with which a builder and two assistants progressed with work on a part-built boat. We were told that a *parical* costs Rs 1,500 and that a boatbuilder earned about Rs 200 each day. On reflection, it seems possible that Rs 1,500 is the cost of materials and that the builders are paid the daily rate.

Size of boat was given as the diameter at the mouth measured in 'feet', the unit being the builder's own foot (*c.* 9 in. or 230 mm). *Parical* at Hogenakal were generally of two sizes: those of 8 'feet' (*c.* 1.9 m) which were used for fishing; and those used for tourists, which could be up to 11 'feet' (*c.* 2.57 m) in diameter. The representative boat measured (Figs 8.4 and 8.5) had a diameter of *c.* 2.24 m (i.e. *c.* 9.6 'feet'). The depth of an average boat was said to be 2 'feet' (0.47 m) and the *parical* measured was approximately this depth.

Raw materials

The *parical*'s framework is made of bamboo (*Bambusa* sp.), in Tamil *mungal*, which comes from sites some 20 km away. Up to 30 bamboo stems are required for an 'average' boat. The 'skin' of the boat was formerly made of ox or buffalo hides – approximately $2\frac{1}{2}$ hides for an 'average' boat. However, about ten years ago a double layer of sewn-together fertiliser bags began to be used. Tar (possibly bitumen) is used as a paying (see Glossary) and to seal the seams in the 'skin'. The lashings used to fasten the 'skin' to the framework are of coir (probably *Cocos nucifera*).

Building the Hogenakal *parical*

The primary framework

The framework is built of 25 to 30 bamboo stems or canes, each some 13 ells/cubits (*c.* 6 m) in length and individually selected for flexibility and

Figure 8.5 Oblique view of the *parical* featured in Fig. 8.4.

'suitability'. Each bamboo stem is split in two across a diameter: some of these are used to form the secondary, reinforcing framing (see below). The remainder are split again into quarters, and these quarters are split tangentially to form thin 'strips' with one plain, worked face and one curved, shiny, outer face. These strips are woven together into an open basketry to form the framework for the bottom and sides of the boat.

Each unit of this primary framing consists of two of these strips laid side by side, and these units are woven across the bottom in three different directions (a tri-axial weave). The first group of paired strips is laid parallel to a diameter of the boat's bottom with all strips extending beyond the intended diameter of the boat. The other two groups are opposed diagonals, crossing the first pairs at an angle of 60 degrees and intersecting each other at 120 degrees (see Figs 8.4, 8.5); these also extend beyond the boat's bottom. A rule of thumb appears to be that, for each 'foot' in diameter, one pair of strips is needed in each of the three groups: thus a boat 11 'feet' in diameter should have a total of 33 pairs.

The builder begins by placing the first group of paired strips on a flat area of ground, curved side down (said to decrease the amount of water trapped and thus reduce the tendency to rot) and some 3 to 4 inches (70 to 100 mm) apart. One group of diagonal pairs, again outer side down and with the units 70 to 100 mm apart, is then woven alternately under and over the first group. This process is then repeated using the other group of diagonal pairs. The result is an open-basketry primary framework with a hexagonal mesh sided *c.* 75 mm (Fig. 8.6), similar to that used elsewhere in Tamil Nadu to make fish traps. The bottom is completed and given structural integrity, and the lowest point of the sides defined, by a circumferential band (in Tamil, *suttu debbai*) of three flat strips woven into the existing lattice (Fig. 8.7).

The sides of the boat are then formed by weaving 20 to 30 adjacent single strips of bamboo circumferentially under and over the protruding ends of the three groups of bamboos forming the bottom. The lowest of these girdling bamboos is set at a known distance from the *suttu debbai*; the uppermost one marks the top of the sides, height of side being judged by eye. As in basket-making, as each of these girdling bamboos is added it moulds the boat to a flat-bottomed bowl shape. These side bamboos are laid with their curved, outer surface inboard, evidently to make the boat more 'passenger-friendly'.

To keep the sides rigid, this structure is strengthened by fastening a stout cylindrical bundle of about 30 thin split bamboos around the inside of the rim (Fig. 8.8). This bundle is, in effect, a gunwale: it protrudes inboard by 60 to 70 mm, and in depth it tapers from *c.* 60 mm to *c.* 30 mm inboard. One or two additional split bamboos are next passed around the girth of the framework, outboard of the rim, and the 'gunwale' bundle is temporarily lashed to them and to the framework with coir rope. The gunwale, the framework and the additional split bamboos are subsequently permanently lashed together by two men working in a systematic fashion from opposite

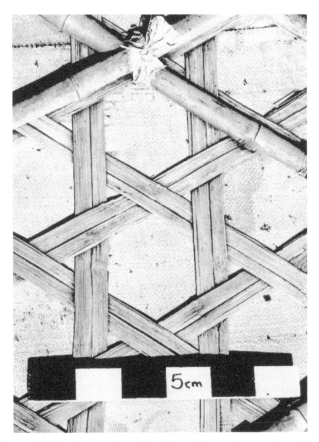

Figure 8.6 Detail of the hexagonal mesh of the primary framework, with part of the
secondary framing visible at the top of the photograph.

sides of the boat. There is one lashing, consisting of two round turns and a
reef knot, in each space between the pairs of bamboos strips. At this stage
these pairs still extend above the rim of the primary framework, and the two
ends of each lashing are left overlong (see Fig. 8.8). By this systematic
lashing sequence, and by beating the bundle of bamboos with the back of a
sickle or a wooden mallet, the boat is given a near-circular shape and a level
rim. It is relevant to note that the *parical* measured had diameters ranging
from 2.19 to 2.27 m (9.4 to 9.7 'feet').

The waterproof cover

The 'skin' is made from twelve plastic fertiliser bags (of a woven material
such as polypropylene). Six of these bags are split open and machine-stitched
together to form a near-square in shape. This is laid flat on the ground, tarred

Figure 8.7 Detail of the *parical* structure from inboard. From the bottom of the photograph: the circumferential band at the perimeter of the boat's bottom; canes woven into the primary framework to form the sides; the gunwale bundle; and the lashings which fasten the bundle and the 'skin' of the boat to the framework. Two elements of the secondary framing can also be seen.

on its upper surface and a second, similarly shaped membrane of six bags is laid on top of it.

The *parical* framework is then placed centrally on this double membrane and the 'skin' is cut by sickle to a rounded shape and to a size which will allow it to cover the bottom and sides of the framework, with an overlap so that its upper edge can be hemmed. The 'skin' is pulled taut and pierced by a pointed bamboo stick to make two holes at the level of the lower edge of the gunwale, adjacent to each lashing fastening the gunwale to the framework. The lashings are passed through these holes from inboard to outboard, the 'skin' is hemmed by rolling up the excess and tucking it into the framework, and each lashing is then led over the gunwale to the inside of the boat. At each lashing station, the inboard bamboo of those forming the gunwale is then prised away from the bundle, using a thin bamboo fid. Each lashing is pulled taut, passed through this gap three times, and the fid removed, thus trapping the lashing within the gunwale bundle (Fig. 8.8). To tighten the 'skin' over the framework, a wooden mallet is used to force the gunwale bundle upwards, thereby taking up any slack in the 'skin'.

Figure 8.8 Fastening the 'skin' to the framework. The builder threads overlong lashings through the 'skin'. A fid has been used to prise the innermost cane away from the gunwale bundle; the lashings will subsequently be trapped within the gap so formed. Note that the ends of the canes forming the primary framework still protrude above the mouth of the *parical*.

Bamboos protruding above the top edge of the sides are then cut off flush with the gunwale, the boat is inverted and the 'skin' covered with hot tar.

The secondary framework

When the tar has hardened, the primary framework is reinforced by three pairs of stout halves of split bamboo (plane surface outboard), which are about 6 cubits/ells (*c.* 2.75 m) in length. These bamboos are positioned on top of the primary framework in the order shown in Figure 8.4: they run from gunwale bundle to gunwale bundle. Bamboos 1 to 3 are interwoven, then 4 to 6 (Figs 8.4, 8.5). Bamboos within each pair are *c.* 0.42 m apart; pairs are set at approximately 60 degrees to each other. We believe, but are not certain, that these bamboos are sprung into place under the bundles, so tensioning the whole structure. They are fastened together at the interstices, and occasionally to the primary framework, with paper-covered wire.

Equipment

A spar about 18 inches long is fastened to the Hogenakal *parical* framing by a short length of line and used as a support by the fisherman when he carries the boat. It can also be used as a mooring spike plunged into the river bank.

A paddle (*thuduppu*) made of a local wood (probably mango – *Mangifera indica*) is used for propulsion (Fig. 8.9). A typical one measured 1.34 m overall, including a handle of 1.00 m which was *c.* 35 mm in diameter. The blade was *c.* 190 mm broad and 10 mm thick.

Propulsion and steering

Parical are launched by being turned over into the water from the carrying position across the fisherman's shoulders. Boats are both propelled and steered by a single paddle (see Fig. 8.9). The fisherman sits or squats on his heels on the framework, facing forward in the leading part of the boat, and uses a variety of strokes: alternating from one side to the other; sometimes sideways; and sometimes using a figure-of-eight stroke. By varying both stroke angle and stroke strength he simultaneously propels and steers the boat. Passengers squat around the periphery.

Figure 8.9 A paddle in use from a *parical* at Hogenakal.

Uses

When used for fishing, each boat has a crew of two: one to paddle and one to fish by throwing a weighted net into the river in a circle. The boat is paddled downstream from the village of Hogenakal, helped by the river flow, to a point above the falls. The boat is carried from there overland to a large pool, below the falls, where fishing takes place. Fish taken include *katla, rohu, kendai, keluthi, valai, mirgal, aranjan* and *jilaby* – the scientific names of these are unknown to us. When returning upstream the crew walk where necessary, carrying boat and fish.

When used for tourists, boats are manned by only one man: six to eight passengers can be carried, depending on size of boat and on river state. Again the boat is taken downstream to a position above the falls and the tourists walk around the falls to view them from below. A round trip takes about one hour, depending on the river flow, and the charge is Rs 30 for each passenger. More time is now spent on tourist trips than on fishing, as the former is more profitable. We were informed that 50 to 60 boats could supply enough fish for the village, the remainder were used in the tourist role.

Despite the great skill used by paddlers, a boat can bump into one of the many rocks protruding from the river bed (Deloche, 1994: 35). Generally only the 'skin' is punctured in such incidents; the framework, being resilient, is undamaged. Small holes and tears in the 'skin' are repaired by the fishermen themselves, using either a tarred patch of fertiliser-bag material or tar alone.

Because of the fragile nature of the 'skin', *parical* cannot be placed right way up on the margins of the river where there are many rocks. They are generally 'parked' upside down (see Fig. 8.2) or may be left for a while leaning against a tree or large rock (see Fig. 8.5).

Discussion

Two boats from Karnataka

The Department of Ethnography at the British Museum has two *harigolu* (the Kannada word equivalent to Tamil *parical*) acquired from Vijayanagara in Karnataka in *c*. 1986 (Figs 8.10, 8.11). An annotated photographic record of the building of these boats is held in the British Museum. The larger of these two (*c*. 3.21 m, i.e. 13 'feet', in diameter and 0.73 m deep) has a plastic 'skin'; the smaller one (*c*. 2.88 m, i.e. 12 'feet', by 0.76 m) has hide. This hide cover was made of two large portions of hide and several smaller pieces, all sewn together to form the required shape; it is now *c*. 5 mm thick. In general these *harigolu* are similar to the Hogenakal *parical*, but they differ from the latter in the following ways:

Figure 8.10 The British Museum's plastic-covered *harigolu*. Note the tertiary framing.

(A) The lowest circumferential band in the primary framework, which in the Hogenakal boats marks the lowest point of the sides, is some way up the sides of the two Vijayanagara boats and consists of only two, rather than three, strips of bamboo.

(B) The gunwale bundles are of whole canes: this seems to be the case in the boats illustrated by Hornell (Fig. 8.3) but is not so in the Hogenakal boats, where split canes are used.

(C) The secondary, reinforcing framework consists of many more canes: 15 in the smaller boat and 18 in the larger, rather than 6. In this respect the Vijayanagara boats are probably like Hornell's Type B (Tungabhadra) boats, which had 'a considerable number' (Hornell, 1946: 95). As in the Hogenakal boats, the secondary framework in the British Museum boats consists of split halves of bamboo extending from gunwale to gunwale, but they intersect near the centre of the boat's bottom in a complex manner and are not fastened together at the interstices.

(D) A most obvious difference is that these two boats have a tertiary, inner framework consisting of split wood of an unknown species, interwoven

254

Figure 8.11 The British Museum's hide-covered *harigolu*. Note the 'egg-shaped' profile. The scale is 0.30 m overall.

in groups of four in a tri-axial weave to produce another hexagonal pattern (Fig. 8.10). The ends of the canes in this tertiary framework appear to have been forced under the gunwale bundles. Although this inner framework would have further reinforced the hull, it seems likely that its principal function was to protect the secondary and primary framework, as well as the 'skin', from damage by cargo and passengers.

(E) While the larger, plastic-covered, Karnataka boat is saucer-shaped in section, with a diameter/height ratio of 4.41, and the Hogenakal boats are bowl-shaped (d/h = 4.67), the hide-covered boat is shaped more like one end of an egg (d/h = 3.79), with much less area of flat bottom (Fig. 8.12); in this respect it may resemble Hornell's Kurnul boat (Type C), which had a 'sharply convex' shape (1946: 95).

(F) The 'skin' of the British Museum boats – both plastic and hide – is not fastened to the framework using the ends of the many gunwale binding ropes, as in the Hogenakal *parical*, but by a single circumferential, spiral coir lashing.

Indian hide boats during the twentieth century

Apart from the plastic 'skin' on the larger boat, the British Museum's boats from Vijayanagara appear to be similar to the larger *parical* (Type B) that Hornell saw in Karnataka in the 1930s and summarily described (1946: 94–5). Hornell did not mention a tertiary framework, but his descriptions of the boats he observed are not as clear or as comprehensive as they might be, and it is conceivable that at least some of the larger boats of his time did have this additional framework.

255

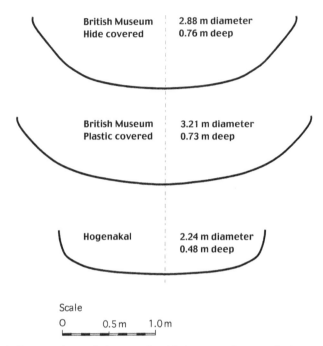

British Museum   2.88 m diameter
Hide covered   0.76 m deep

British Museum   3.21 m diameter
Plastic covered   0.73 m deep

Hogenakal   2.24 m diameter
                   0.48 m deep

Scale
0   0.5 m   1.0 m

Figure 8.12 Cross-sections of the two British Museum boats and the representative Hogenakal boat.

In general terms, the Hogenakal *parical* of today is also similar to the British Museum Vijayanagara boats but has a simplified structure. It may be that early Tamil hide boats of a similar size and function to today's Hogenakal *parical* never had complex framing. Or it may be that, in the 1980s, when plastic 'skins' evidently began to replace the traditional hide cover and tourists began to increase in numbers, the framework of the Hogenakal *parical* was simplified by omitting the tertiary layer and by reducing the number of bamboos in the secondary layer, but fastening the remaining ones together at the interstices.

The fishing *parical*, as described to us by fishermen, appears to be similar in size to Hornell's Type A (Kaveri); the boats now used at Hogenakal for tourists are about the same size as Hornell's Type B (Tungabhadra).

The virtues of the parical

The hide boat of southern India seems to have been developed to match its river environment and the roles required of it, sometime before (probably well before) the late fourteenth century when this type of craft was first noted. Such boats can be built with simple tools and techniques from readily available bamboo, coir and hides. They can be used in a variety of roles on

256

rivers that vary markedly, often during the course of one day, in their courses, depths and speed of flow; and they can readily be carried from river to river. The Tamil Nadu hide boat's role as a river fishing boat has continued into the twenty-first century. Its cargo-carrying role seems to have declined and then been terminated as better roads were built. Its use as a river ferry began to be curtailed with the building of bridges, but at Hogenakal it continues to be used in this role, as do similar hide boats at Vijayanagara/Hampi in Karnataka.

When asked why the *parical* was circular in shape, several informants told us that if it were to be 'boat-shaped' it would be quickly overturned in the fast-flowing, rock-strewn River Kaveri. The *parical*'s round shape allows it to be propelled and steered from any position and ensures good manoeuvrability. A circular form is also the simplest way of making a basket-style framework. A lightweight yet resilient structure allows it to be used in shallow water and also to transport relatively great loads, yet, when empty, it can be carried by one man. Simplicity, lightness, cheapness and fitness for purpose are the hallmarks of the South Indian circular hide boat.

Future research

Some of the Hogenakal fishermen thought that *parical* were still being used elsewhere on the River Kaveri, but this remains to be confirmed. Hide boats were certainly used in the late 1990s on the River Tungabhadra in Karnataka. It may be that they are still used at other sites in Karnataka and also in Andhra Pradesh; further fieldwork is required here.

This chapter is based on fieldwork undertaken during a fleeting visit to Hogenakal and a rapid recording of one boat, followed by necessarily short interviews with two or three fishermen and boatbuilders. It is believed that this description of the structure and the use of the Hoganakal *parical* is accurate in all essentials. Nevertheless, we consider that a second, longer, visit to Hogenakal would be desirable at some future date.

There is a Tamil Nadu boat similar to, but smaller than, the Hogenakal *parical* in the ISCA collection at Lowestoft (No. 204). This has been documented by the Centre for Maritime Archaeology, University of Southampton. Two Karnataka *harigolu* similar to the pair in the British Museum are in the Von Portheim Stiftung, the ethnographic museum in Heidelberg. There may be further examples of South Indian hide boats in other European institutions: these need to be traced and documented.

9

A HYDRODYNAMIC EVALUATION OF
FOUR TYPES OF BOAT

Colin Palmer

Four boat types, the *vattai, vallam, patia* and *barki*, are compared in this chapter from the point of view of their naval architecture. Figure 9.1 shows their hull profiles, drawn to a common scale so that overall size and proportion may be compared. Figure 9.2 presents perspective lines plans of the hulls, in this case adjusted to similar sizes on the paper for ease of visual comparison. The overall dimensions of each hull, as well as a number of key parameters calculated from their lines, are listed in Table 9.1 below.

For the *vattai, patia* and *barki*, this information was obtained by taking measurements of cross-sections and overall length from the detailed drawings presented in Chapters 7, 3 and 2 respectively. The *vallam* had not been recorded in such detail, so a combination of field measurements and scaling from photographs was used to create the measurements for a 'typical' large *vallam*.

The measurements were used as underlying reference markers for the preparation of a lines plan using the industry standard MaxSurf lines drawing computer package. This program uses an interactive spline fitting approach to the preparation of hull lines. This means that it reflects the process of wooden boatbuilding where the hull planks act as 'splines' that fair the hull during the construction process. Thus, although there was inevitably some loss of accuracy in the transfer of measurements from the field to the detailed drawing and then to the lines drawing, the natural fairing process of the MaxSurf program ensured that the final lines drawing was fair and a practical representation of the original.

As can be seen from Figure 9.1, the four boats differ greatly, not only in overall length but also in profile shape and proportion. At one extreme is the *vattai*, which is 11.5 m long on the waterline and has a high sweeping sheer at the bow. At the other extreme, the *barki* is half the length of the *vattai* and has a gentle sheer which is actually lower at the bow than the stern. There are similarly large differences in the hull cross-sections. In view of these extremes, the analysis which follows provides an individual description of

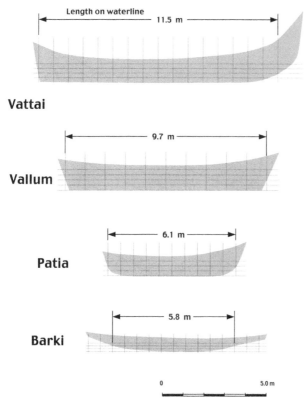

Figure 9.1 Comparison of hull profiles drawn to the same scale. Length of waterline indicated for draught equal to 60 per cent of hull depth amidships.

each boat and then makes use of non-dimensional parameters to compare their technical characteristics.

Resistance prediction

The calm water resistance of the hulls was predicted using a parametric method developed in recent years for the resistance prediction of sailing yachts (Keuning and Sonnenberg, 1996). It is a technique derived from the results of a large number of model tests on a wide range of sailing yacht hull forms. As such, it can be expected to give reasonable accuracy in the prediction of the resistance of small boat hulls of the type under consideration. The main possibility for error is that the more 'ship-like' midship sections and parallel midbodies of the *vattai* and *vallam* are not found in sailing yachts, so these forms would not have been explicitly covered in the underlying tank test models. However, the prediction method is based upon hull coefficients and all the types considered in this paper fall within the range where satisfactory

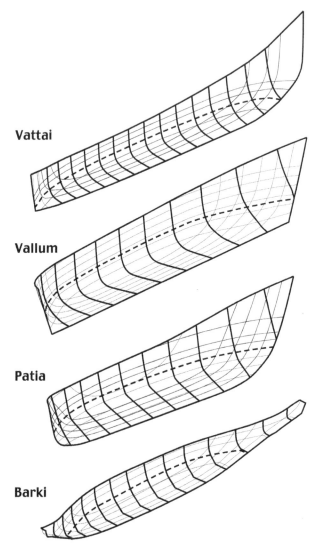

Figure 9.2 Perspective lines plans of the four types to allow comparison of the hull forms (not to scale).

accuracy may be expected from the parametric model. It should also be noted that the model only predicts residuary resistance (i.e. primarily the wavemaking resistance). The frictional resistance is predicted from the wetted surface area and a standard skin friction coefficient for rough hull surface finish.

In practice, traditional boats propelled by sail and oar operate almost exclusively in the speed range below a Froude number (Fn) of 0.35, which means that they are operating in conditions where the frictional resistance is a major, if not dominant, resistance component. (Specialist technical naval architectural terms such as Froude number are defined in Annex I to this chapter). Therefore, for practical purposes, small errors in the prediction of the residuary resistance are unlikely to have a significant effect on the understanding of the performance potential of the boats.

It is important to note that, for small boats in particular, the calm water resistance is only a baseline guide to performance: under most practical conditions at sea there will be a significant component of added resistance due to waves. This effect is most pronounced in waves approaching from forward of the beam and, in head seas, it can result in an increase in resistance of 50 per cent or more. Added resistance in waves is an extremely complex subject; its magnitude depends upon factors that relate not only to the shape of a boat's hull but also to its distribution of mass, the relationship between the frequency of the waves and the length of the boat, and the height of the waves. Recent studies (Gerritsma *et al.*, 1993; Sclavounos and Nakos, 1993) have investigated the hull form parameters that influence added resistance in waves: the two parameters that have the greatest effect are the length-to-beam ratio (L/B) and the separation between the longitudinal centre of buoyancy (LCB) and the longitudinal centre of flotation (LCF). Added resistance reduces with increasing L/B ratio and with increasing separation between LCB and LCF. For example, the added resistance in waves of a hull with an L/B ratio of 2.8 is more than twice that of a hull with an L/B ratio of 4. A separation of 7 percentage points (e.g. LCB at 53 per cent aft, LCF 60 per cent aft) halves the added resistance when compared to a hull form in which they are coincident.

The radius of gyration (longitudinal moment of inertia) also has a moderate to strong effect. This is however determined mainly by the load distribution within the hull rather than the hull form. As such, it is a variable under the control of the crew and is not included in this discussion.

The *vattai*

Hull form

The *vattai* is a double-ended hull form with a very marked sheer. The sheer forward is extreme and results in the stem-head height being more than twice the midships hull depth. It is difficult to understand how this extreme form can be interpreted in performance-related terms. Certainly, a rising sheer forward is useful in keeping a boat dry in a seaway, but the extreme shape of the *vattai* bow is very unlikely to be superior in this respect to a more modest shape. It is also noticeable that smaller *vattai* and the related types found in the same region do not have the same striking high bow shape. In combination with the high bow, the *vattai* has a curved forefoot, and in some cases – though not in the example illustrated – rocker (see Glossary) is introduced to the keel some way back from the bow, further increasing the curvature in the region of the forefoot. In performance terms, this shape may exhibit a tendency to slamming in waves.

At a draught of 60 per cent of the midship hull depth (the international standard for comparison), the *vattai* hull has a waterline length-to-beam ratio of 6.65 and a beam-to-draught ratio of 2.88. These describe a slender and shallow hull form. The hull depth amidships is just 9 per cent of the waterline length. This rather low ratio is compensated by the rising sheer at the ends, so the safety of the boat at sea is unlikely to be compromised unless it is heavily overloaded.

The *vattai* midship section has no rise of floor and a moderate bilge radius. The topsides are vertical and flare is only introduced at the ends of the hull. The midship section coefficient (Cm) of 0.90 is unusually high for a small wooden boat. It is a direct result of designing the hull with a small bilge radius and no rise of floor in the midship sections. Another feature that is unusual is the extent of the parallel midbody – normally a feature found in large cargo vessels but not in small boats. This again arises directly from the design method, which uses a number of identical frames either side of midships. The parallel midbody results in a high waterplane area coefficient (Cw) of 0.82 and a prismatic coefficient (Cp) of 0.72.

Resistance

The predicted calm water resistance curve for the *vattai* is shown in Figure 9.3. A Froude number of 0.35 is equivalent to a speed of almost 3.75 m/sec (7.3 knots). At that speed the frictional resistance is 85 kg and the residuary resistance is 74 kg.

Figure 9.4 shows a comparison of the resistance per tonne of displacement for all the hull forms being described. It indicates that even at typical operating Froude numbers (up to 0.35) the differences between the hull forms are substantial and they become very marked at high speeds. The *vattai* has the least residuary resistance per tonne of all over the complete speed range. This suggests that, under rare conditions of strong winds in a favourable direction, the *vattai* has the greatest potential for achieving high speeds under sail. (The important influence of balance boards on this potential is discussed below.)

The high L/B ratio of the *vattai* is good for operation in waves because it will result in low added resistance, as argued above. However, the lack of separation (less than one percentage point) between the LCB and LCF is not beneficial in this respect.

Stability

The firm-bilged, shallow-draught hull midship section will produce a boat that has high form stability, but because of the vertical sides this will be

CONSTANTS

LWL	11.50 metres
BWL	1.73 metres
Friction coefficient	0.005 Rough fishing boat hull
Water density	1.03 kg/cu m

HULL CALCULATIONS

Displacement	7.99 tonnes
Vol Disp	7.79 cu m
WPA	16.26 sq m
WSA	23.52 sq m
Cp	0.72

LCB, LCF

LCF	48.3	% LWL
LCB	48.8	% LWL

RESISTANCE CALCULATIONS

Fn	Speed m/sec	Residuary Resistance Kg	Friction Resistance Kg	Total Resistance kg
0.20	2.12	-2	28	25
0.25	2.66	14	43	57
0.30	3.19	26	62	89
0.35	3.72	74	85	159
0.40	4.25	163	111	274
0.45	4.78	295	140	436

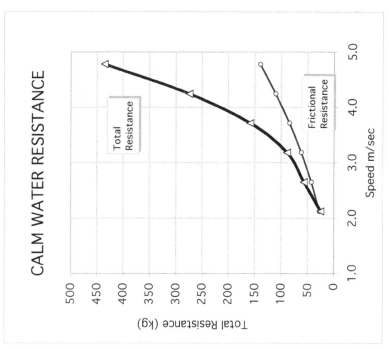

CALM WATER RESISTANCE

Figure 9.3 Results of spreadsheet calculations of hull resistance of a *vattai* at 0.6 m draught, using the Delft resistance algorithm (Keuning and Sonnenberg, 1996).

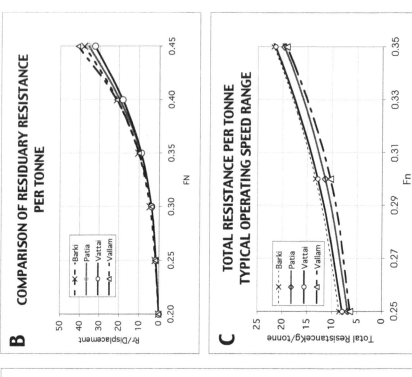

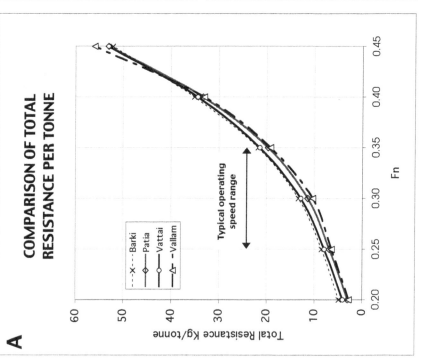

Figure 9.4 Comparisons of calm water resistance results obtained from calculations: A) residuary resistance plus skin friction over a wide speed range; B) residuary resistance over the same speed range; C) enlargement of part of A.

reduced when heavily loaded. From the point of view of the crew, the boat will feel reasonably stiff under normal operating conditions. One person moving from the centreline to the gunwale will induce a heel angle of just over 1.5 degrees, which will be perceptible but not alarming.

Sailing performance

The very low aspect ratio (length-to-draught ratio) of the *vattai* hull means that the hull alone will be very ineffective in producing the lateral forces required for reaching or close-hauled sailing. However, with the wind astern or on the quarter the requirement for lateral forces more or less disappears. In these conditions, the rising bow and fine stern fairing into what is in effect a deadwood in front of the rudder should give a hull which is stable and controllable even in following seas.

For close-hauled sailing the fishermen use one or two (sometimes even three) leeboards on their boats. These boards not only provide the lateral forces required on this point of sail, but, by varying their longitudinal position, it is also possible to change the balance between hull and rig lateral forces. This allows the boats to be controlled with what are otherwise rather small rudders in proportion to the size of the boats (due to the rudder depth being constrained by the shallow draught).

The potential sailing performance is further enhanced by the use of balance boards: their effect is twofold:

- they provide a means of placing weights (crew members, sandbags or large stones) to windward to balance the heeling forces from the sails; and
- they increase the width of the shroud base for the rig.

These two functions are interdependent to the extent that the increase in the shroud base allows more sail to be set, which in turn requires greater balancing forces. The wider shroud base is effective because it reduces the compression forces in the mast and the tensions in the standing rigging. This will enable the rig to 'stand' against stronger winds or when sheeted for close-hauled sailing. By attaching the standing rigging of the fore mast and mizzen mast to the balance board, the need for backstay and forestay (respectively) is removed as the forces are triangulated.

The balance boards provide a substantial increase in the power to carry sail. Table 9.1 indicates that, at a typical vertical centre of gravity height of half the hull depth, the righting moment per degree of heel is 38 kg.m. Two men on a balance board can provide an additional righting moment of approximately 400 kg.m (depending on their weight and the length of the balance board). Open working boats are handled with care and they generally appear to sail with a heel angle within the range of 5 to 10 degrees. For the

Table 9.1 Dimensions and parameters of four South Asian boat types

Parameter	Units	Boat type			
		Vattai	Vallam	Patia	Barki
LWL (L)	Metres	11.50	9.90	6.19	6.36
Draught (T)	Metres	0.60	0.72	0.49	0.27
BWL (B)	Metres	1.73	2.62	1.26	1.06
Displacement (Δ)	Tonnes	7.99	12.60	1.78	0.93
Midship Depth (D)	Metres	1.00	1.20	0.80	0.45
LCB	% LWL aft	48.80	50.70	49.30	49.00
LCF	% LWL aft	48.30	50.50	49.80	49.50
WSA	Sq. metres	23.50	28.45	8.28	6.34
Cp		0.72	0.71	0.70	0.66
Cw		0.82	0.78	0.73	0.79
Cm		0.90	0.94	0.65	0.76
BMt	Metres	0.43	0.76	0.32	0.42
GMt	Metres	0.27	0.55	0.24	0.36
KG (half depth)	Metres	0.50	0.60	0.40	0.23
Righting Mmt/degree	Kg metres	37.88	121.51	7.38	5.91
L/B		6.65	3.78	4.90	6.03
B/T		2.88	3.64	2.59	3.91
D/L		0.09	0.12	0.13	0.07
$\Delta/0.001(L)^3$		5.25	12.99	7.49	3.62
$WSA/(\Delta^{2/3})$		5.97	5.34	5.74	6.76
BMt/B		0.25	0.29	0.26	0.40

Notes
LWL = Length of vessel at waterline
BWL = Breadth of vessel at waterline
BMt = Distance, centre of buoyancy to metacentre
GMt = Distance, centre of mass to metacentre
KG = Distance, 'keel' to centre of mass
Other parameters are defined in Annexes I and II to Chapter 9

vattai this would correspond to a righting moment of 190 to 380 kg.m. Thus the use of the balance board increases the heeling moment at 5 degrees by 200 per cent and at 10 degrees by 100 per cent. Compared to a boat without a balance board, this will enable a significant increase in the sail area that can be carried, or allow any given area to be carried in stronger winds. The effects of the wider shroud base, as described above, enhance this capability.

Use of one, two or three sails

The materials available to the boatmen (bamboo spars and coir/nylon rigging) mean that the rigs have considerable flexibility. The potential height of the rig relative to the shroud base will be limited by deflection rather than the ultimate strength of the materials. As a consequence, the most practical means of increasing sail area is to increase the number of masts rather than increase the size of any one sail. The use of multiple masts also confers other advantages, particularly in a culture where reefing appears to be unknown:

- the sail area can be readily varied to suit changing wind conditions
- the centre of effort of the rig is kept low
- the balance of the hull/rig combination can be varied
- the forces from the rigging are distributed throughout the hull

When combined with the use of up to three leeboards, which can be placed at a number of longitudinal locations, the overall result is a very versatile working rig with the potential for very good all-round performance. The main limitation is that a multi-masted rig is less efficient when sailing close hauled. The crew appear to recognise this to the extent that they generally use a maximum of two sails set on the fore and main masts, when working close to the wind. We were told that they only use three sails in very light winds or when trawling. In strong winds they only use one sail, set on the main mast.

Sail shape

The high-peaked settee-shaped sail has a spar over most of the leading edge. This will have some detrimental effect on the flow over the sail but also the beneficial effects of providing a straight luff and controlling the sail twist. These effects can be expected to provide a powerful reaching sail and one that can be set close hauled. The high peak of the sail is curved and flexes considerably, which will provide a degree of automatic load shedding in gusts. The fishermen use bowlines on the short unsupported luff, which will further improve the close-winded ability of the rig.

Sail area and power to carry sail

No direct measurements were taken of *vattai* sails, but examination of a number of photographs indicates that the length of the main mast above the sheerline is typically 50 per cent of the overall length of the boat. It was also possible to deduce rough sail proportions from a photograph. This gave the following ratios for the lengths of sail sides relative to mast length:

Yard	1.55
Leach	1.55
Luff	0.36
Foot	0.91

This gives a formula for sail area of:

$$\text{Sail area} = (\text{mast height})^2 \times (0.91 \times (0.36+1.55)/2)$$

This simplifies to:

$$\text{Sail area} = 0.87 \times (\text{mast height})^2$$

Thus, for an overall hull length of 12.5 m, the area of the mainsail will be about 34 square metres. The foremast is typically 88 per cent of the main mast length and, assuming that the foresail is in the same ratio to its mast, the area of the foresail will be around 26 square metres. This gives a total working sail area of 60 square metres. The centre of effort of the rig will be at around half the mast height – about 3 metres above the sheer and thus around 4 metres above the hull centre of lateral resistance. In an apparent wind of 15 knots (typical reaching conditions) the lateral force from the sails will be around 200 kg; thus the heeling moment will be 800 kg.m. The combined righting moment of the hull at 5 degrees with two men on the balance board is comparable at almost 600 kg.m. (Although this figure is lower than the calculated heeling moment, in practice the sails and spars can be expected to deflect under load, thus reducing the forces and moments as compared to theoretical calculations.) As the lateral and driving forces are roughly equal on the fastest point of sailing (a beam reaching course), the driving force is thus in the region of 200 kg. This is the same as the calm water resistance at 4 m/sec (Fig. 9.3). This analysis therefore confirms that the *vattai* can achieve speeds of around 8 knots (4 m/sec) in calm water before stability becomes a limiting factor.

The *vallam*

Hull form

The *vallam* is also a double-ended hull form but, unlike the *vattai*, it has a distinct plank-keel (rather than a foundation plank). The hull has no rocker because it is constructed from a straight keel with straight stem and stern posts joined at an angle to the keel. The hull has a sharp forefoot and similarly shaped afterbody. The *vallam* has a more gentle sheer than the *vattai* but greater midships hull depth. The fine entry and marked forefoot of the *vallam* can be expected to minimise any tendency to slamming in waves and produce a forward position for the centre of lateral resistance.

At a typical operating draught of 0.72 m (60 per cent of hull depth) the *vallam* hull has a waterline length-to-beam ratio of 3.78 and beam-to-draught ratio of 3.64. These parameters describe a relatively bluff and shallow hull form. The hull depth amidships is 12 per cent of the waterline length. This ratio, coupled with the gently rising sheer, results in a hull form with more than adequate freeboard.

Like the *vattai*, the *vallam* midship section has no rise of floor and a moderate bilge radius. The topsides are vertical and flare is only introduced at the ends of the hull. The midship section coefficient (Cm) of 0.94 is unusually high for a small wooden boat and arises for the same reasons as in the *vattai*. However, the *vallam* has less parallel midbody than the *vattai*, which results in a slightly lower value for the waterplane area coefficient (Cw) of 0.78. The prismatic coefficient (Cp) is 0.71, which is very similar to that in the *vattai*.

Resistance

The predicted calm water resistance of the *vallam* is compared to the other boats in Figure 9.4A. The residuary resistance per tonne of the *vallam* is the highest of all the four hulls over the complete speed range (Fig. 9.4B). This is primarily due to the lower length-to-beam ratio and higher displacement-to-length ratio. However, these same ratios also mean that the *vallam* has less wetted area per tonne of displacement, so the total resistance per tonne is the lowest of all the hulls in the normal operating range of 0.25<Fn<0.35 (see Fig. 9.4C, which is an enlarged section from Fig. 9.4A).

When operating in waves, the *vallam* can be expected to be a poor performer in terms of added resistance. It has a low L/B ratio and almost no separation between the positions of the LCB and LCF.

Many *vallam* are being mechanised, although the hull is not a form that is ideally suited to mechanisation owing to the fine afterbody, which could lead to squatting (see Glossary) with high power levels. However, the relatively modest power levels currently used appear to give a satisfactory performance and *vallam* were not observed to squat excessively. The fine afterbody also provides a clean inflow to the propeller. The fact that the fishermen continue to use both sail and power is a further indication of the modest levels of power.

As in the *vattai*, the firm-bilged, shallow-draught hull midship section gives a boat that has high initial stability. Although the *vallam* is shorter than the *vattai*, its substantially greater beam means it has a higher displacement and greater form stability. As a result, one man moving from the centreline to the gunwale will induce a heel angle of about 0.7 degrees – which is half the angle for an equivalent movement on the *vattai*. This will give the *vallam* a 'stiff' feel.

Sailing

The *vallam* is a much less slender hull form than the *vattai* and has a significantly higher displacement-to-length ratio. The combination of these two factors means that it has a lower maximum speed potential. However, this limitation will be offset to some extent by the much greater power to carry sail that results from the *vallam*'s greater beam and displacement.

Some *vallam* were observed to use short bamboo poles as 'balance boards', but this practice was far from universal. Perhaps experience has shown that the gains from the added stability that they provide are not as worthwhile as in the more slender *vattai*.

Like the *vattai*, the *vallam* use leeboards even though the presence of a plank-keel may be expected to result in a hull with a slightly superior ability to generate the lateral forces required for close-hauled sailing. However, as tank tests have shown (Fig. 9.5; see also Palmer, 1990: 80), the keel depth below the *vallam* hull would need to be considerably greater than it now is before windward performance became possible.

The *patia*

Hull form

The *patia* is another double-ended hull form, with a plank-keel which is not externally prominent. The reverse-clinker method of construction results in a distinctive midship section with a steep rise of floor and marked turn of bilge. The hull has no rocker, but at the ends the profile curves up in a continuous curve to the stem head and top of the stern post. The *patia* has a gentle sheer with more freeboard forward than aft. The V sections will minimise any tendency to slamming in waves.

The topsides have slight flare amidships which increases towards the ends. The midship section coefficient (Cm) of 0.65 is low for a working boat and close to that expected for a yacht. The hull is full at the ends, which is reflected in the

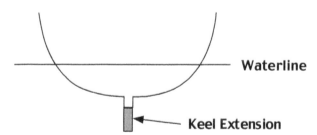

Figure 9.5 Diagram indicating the keel extension required to increase hull draught by *c*. 50 per cent and thus provide a *vallam*-type hull with a modest windward sailing potential. Results from tank tests (Palmer, 1990).

waterplane area coefficient (Cw) of 0.73 and prismatic coefficient (Cp) of 0.70. At a typical operating draught of 0.48 m (60 per cent of hull depth) the *patia* hull has a waterline length-to-beam ratio of 4.90 and beam-to-draught ratio of 2.59. These describe a relatively slender and deep hull form. The midships depth is 13 per cent of the length, which gives ample freeboard for safety at sea.

Resistance

The predicted calm water resistance of the *patia* is compared to the other boats in Figure 9.4. The residuary resistance per tonne is second only to the *vattai* at low speeds but rises more steeply, mainly because of the higher displacement/length ratio of the *patia*. The total resistance per tonne is similar to the *vallam*, a result that is influenced by the relatively low specific wetted area of the *patia* hull form.

When operating in waves, the *patia* can only be expected to be an average performer in terms of added resistance. It has a high length-to-beam ratio, which is beneficial, but almost no separation between the positions of the LCB and LCF. When compared to the *vattai* and *vallam*, the high rise of floor and deep midship section will give a boat that has low to moderate initial stability, particularly when lightly loaded. The effect of one of the crew moving from the centreline to the gunwale will be to induce a heel angle of almost 6 degrees – which is much greater than on the *vallam* and *vattai*. This is in part due to the low displacement of the hull (1.78 tonnes as compared to 12 to 10 tonnes for the *vallam* and *vattai*), but it is also a result of the lower form stability of the hull.

Sailing

The low stability of the *patia* will place severe limits on its power to carry sail and hence on its speed potential under sail. However, the deeply V-shaped hull form will be capable of producing a moderate level of lateral forces, so the boat is probably able to sail under control on a close reach although its windward ability will be very limited.

The sail area of the *patia* is approximately 11 square metres, which is 1.30 times the hull wetted surface area. The equivalent figure for the *vattai* is 2.6. This indicates that the *patia* will be a significantly slower boat under sail. The sail area is most likely limited by the low stability of the hull form.

The *barki*

Hull form

Like the other three boats, the *barki* is a double-ended hull form: in all other respects it is very different. It has a midship section that approaches

271

a semicircular shape and has curved sections over the full length of the hull. The hull also has considerable rocker, with the underside of the keel blending into the prominent *goloi* at either end. The *barki* has a gentle sheer with more freeboard aft than forward – a common feature of the riverboats of Bangladesh. The rounded sections would give a tendency to slam in waves, but, since the boat is used almost exclusively in the calm upper reaches of inland waterways, this is not a significant limitation. For the same reason, the likely added resistance characteristics will not be examined.

The flare of the midship section is carried through towards the ends. The midship section coefficient (Cm) of 0.76 reflects the rounded midship section. The hull is full at the ends so the waterplane area coefficient (Cw) has a high value of 0.79. The prismatic coefficient (Cp) of 0.66 is slightly lower than in the other three boats.

At a typical operating draught of 0.27 m the *barki* hull has a waterline length-to-beam ratio of 6.03 and beam-to-draught ratio of 3.91. These describe a moderately slender and shallow hull form. The midships depth is only 7 per cent of the length, but this low value is not significant for a boat that operates in calm water: it simply sets a limit on the amount of cargo that can be carried.

Resistance

The predicted calm water resistance of the *barki* is compared to the other boats in Figure 9.4. The residuary resistance per tonne is the highest of all at low to moderate speeds but becomes less than the *vallam* at higher speeds. The *barki*'s total resistance per tonne is high at low to moderate speeds, a result that is influenced by the relatively high specific wetted area. At first sight this high wetted surface coefficient may appear surprising, since the sections of the *barki* are close to the 'minimum wetted area' semicircular shape. However, scale effects complicate a comparison which is based upon displacement. Consequently, even in the case of boats of the same proportions and shape, if one has half the displacement of the other, the wetted surface per unit displacement will be 26 per cent higher.

The rounded midship section might be expected to result in a boat that has only moderate initial stability, but the ratio of the BM to beam (BM/B) of the hull is the highest of all the boats. This is most likely a result of having the highest beam-to-draught ratio and lowest displacement-to-length ratio. The effect of one of the crew moving from the centreline to the gunwale will be to induce a heel angle of about 5.5 degrees – more or less the same angle as for the *patia*, despite the *barki* having almost half the displacement of the *patia*.

Conclusions

This chapter has used the tools of conventional naval architecture to interpret the performance characteristics of four different boats and to attempt to understand the operational implications of these characteristics. The approach sheds light on factors such as the maximum speed potential in calm water and waves, load-carrying capability, seaworthiness and stability. It is helpful in explaining what a boat can do and what its ultimate limits are, although this does not mean that these capabilities are necessarily fully exploited in practical operation. The approach is silent on the question of why boats have come to be shaped as they are.

Specific further reading

Barnaby, K.C., 1967. *Basic Naval Architecture*. Hutchinson.
Claughton, A., Wellicome, J. and Shenoi, A. (eds), 1998. *Sailing Yacht Design: Theory*. Longman.
Garrett, R., 1996. *The Symmetry of Sailing*. Sheridan House.

ANNEX I: GLOSSARY OF NAVAL ARCHITECTURAL TERMS

Skin Friction Coefficient (Cf)

The coefficient used in the calculation of the skin friction resistance (Rf). For hydrodynamically smooth surfaces this coefficient is a function of the speed of the vessel. However, for hydrodynamically rough surfaces (which includes those found on almost all traditional boat hulls) the coefficient tends to a constant value of approximately 0.005 (Hoerner, 1965: 5.1).

Midship Area Coefficient (Cm)

The ratio of the immersed midship area to that of the circumscribing rectangle. It is a measure of the fullness of a hull form. Cm can vary over a much wider range than Cp or Cw. For slender, light displacement yachts it may be little more than 0.5, whereas for some fishing boats it can exceed 0.95. The effect of Cm on resistance is complex and depends very strongly on other associated parameters. A typical modern fishing-boat hull may combine a high value of Cm with an optimum value of Cp and a low Cw to give a low resistance form. At the typical operating speed of powered fishing boats, resistance shows a general trend to decrease with increasing Cm (Doust, 1965: 134).

Prismatic Coefficient (Cp)

A measure of the 'fullness' of the hull. It is the ratio of the displaced volume to a volume equivalent to a hull with a constant midship area cross-section. For most small boat hull forms, Cp lies in the range between 0.45 and 0.75. Cp has a strong effect on the residuary resistance and the optimum value varies with speed, with a trend towards increasing with speed.

Waterplane Area Coefficient (Cw)

A measure of the 'fullness' of the flotation waterplane. It is the ratio of the waterplane area to the area of the circumscribing rectangle. For most small boat hull forms, Cw lies in the range from 0.6 to 0.8. Like Cp, Cw has an influence on the residuary resistance, but in a more complex manner. Generally speaking, a hull with fine forward sections (which will tend to be associated with a low value of Cw) will have lower residuary resistance at high speeds.

Cw also has a weak influence on stability. Increasing Cw at constant displacement will increase the stability of a hull. However, an increase in beam (which might reduce Cw) is generally a better way to increase stability.

Froude Number (Fn)

The ratio of the boat speed to the square root of the waterline length. The Froude number is used as a basis for a non-dimensional comparison of residuary (wavemaking) resistance because the speed of a wave is proportional to the square root of its length.

Longitudinal Position of the Centre of Buoyancy (LCB)

The longitudinal position (normally measured as a percentage of the waterline length aft of the buoyancy bow) of the centre of buoyancy of a hull. The LCB position affects the wavemaking resistance of a hull in a complex manner. The overall trend is for the optimum LCB position to move aft with increasing speed. It is almost always in the range between 45 and 60 per cent of the waterline length from the bow.

Longitudinal Position of the Centre of Flotation (LCF)

The longitudinal position (normally measured as a percentage of the waterline length aft of the flotation bow) of the centre of area of the flotation waterplane. The LCF position has a weak effect on the wavemaking resistance of a hull. However, the separation between LCF and LCB has a strong influence on added resistance in waves. The larger the separation, the lower the added resistance.

Height of Metacentre/Beam Ratio (BMt/B)

A measure of the form stability of a hull.

Beam/Draught Ratio (B/T)

The ratio of the waterline beam to the hull draught. This is a measure of the form stability of the hull. A high value of B/T is associated with high form stability.

Depth/Length Ratio (D/L)

The ratio of the hull depth amidships to the waterline length. This ratio provides a measure of the freeboard of a hull and hence an indication of its ultimate seaworthiness and load-carrying ability.

Length/Beam Ratio (L/B)

The ratio of the waterline length to the waterline beam. This is a measure of the slenderness of the hull. Generally speaking, the higher the length-to-beam ratio, the lower the resistance of the hull in calm water and in waves.

Wetted Surface Ratio (WSA/$V^{2/3}$)

A measure of the wetted area of the hull as compared to its volume of displacement.

Displacement/Length Ratio (V/0.001(L)^3)

A measure of the hull 'loading' – the amount of displacement compared to the length of the hull.

Residuary Resistance (Rr)

The components of calm water resistance other than frictional resistance. Rr is primarily resistance due to wavemaking.

Total Resistance (Rt)

The sum of the frictional and residuary resistance.

Wetted Surface Area (WSA)

The area of the immersed surfaces of the hull. Used in the calculation of the frictional resistance.

ANNEX II: STABILITY

Boats gain stability owing to the movement of the centre of buoyancy as they heel. The buoyancy force acts vertically, so the force it exerts crosses the inclined centreline of the hull (Fig. 9.6). The point where it crosses is referred to as the metacentre (M). For small angles of heel (up to 10 to 15 degrees) the position of the metacentre is more or less fixed. The vertical separation (BMt) of the centre of buoyancy (B) and the metacentre (M) is a primary determinant of the form hull stability of the hull – the stability due to its shape. The suffix 't' is used to indicate the transverse direction: a longitudinal BM can also be calculated but is of much less practical significance.

While BMt is an important measure of stability, it is also necessary to take account of the vertical position of the centre of mass (G), because the righting moment produced by a hull is a function of the displacement (Δ), and the separation (GMt) of the centre of mass and the metacentre. As the centre of mass is above the centre of buoyancy in many small craft, GMt is almost always less than BMt.

Stability is a very difficult parameter to compare between different boat types. For example, a large ocean-going liner may actually have a negative GMt yet be perfectly safe. As it heels the metacentric height actually rises

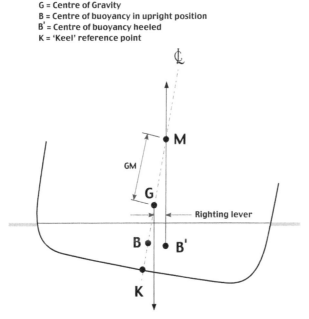

Figure 9.6 Diagram showing the location and directions of action of the forces that determine the transverse stability of a boat.

very slightly, but sufficiently to make GMt positive and hence create a positive righting moment. The high freeboard of the liner, and its large size relative to the waves in which it operates, ensure its ultimate safety.

For smaller boats the effects of differing levels of stability requirements are extremely complex. They influence the movements that the crew members can make, the amount and position of the cargo that can be carried and, perhaps most significantly, the power to carry sail. What may on paper appear to be inadequate stability may in practice be adequate because of the skill of the crew. On occasion, narrow logboats probably operate with negative 'paper' stability but are balanced by the actions of the crew members. A similar effect occurs with rowing sculls and sailboards.

For these reasons it is only possible to comment in very general terms on the stability of a particular hull shape relative to another. The actual values of parameters such as GMt can be the same for vessels of widely differing sizes and shapes. For small boats the effects of movements of the crew within the boat are important, so the effect of one crew member moving from the centreline to the gunwale has been used as a basis for comparison. The resultant heel angle was calculated (line 15 in Table 9.1) using the righting moment per degree of heel angle.

10

THE WAY AHEAD

Seán McGrail, Lucy Blue, Eric Kentley and Colin Palmer

In five seasons of research the Boats of South Asia project has documented examples of three broad traditions of Indian and Bangladesh working boats: the reverse-clinker boats of Sylhet, West Bengal and Orissa; the frame-first vessels of coastal Tamil Nadu; and the hide boat tradition of the upper River Kaveri. The project archive will be deposited at the Centre for Maritime Archaeology, University of Southampton. We believe that the documentation was as accurate and detailed as was possible within the time available and the constraints of our resources. It is necessary now for others to evaluate our methods, criticise our findings and build upon the foundations we have laid.

Although four boats were recorded in detail and two in part, and countless others examined on foreshores and in boatyards, there are gaps in our documentation. We were, for example, never able to see all phases of the building process. In particular we never observed the design stage nor the initial building phase, and our knowledge of these is based on interviews and on *ad hoc* demonstrations. Furthermore, although we went to sea in a *patia* and a *vattai* and were afloat on a river in a *barki* and a *parical*, such voyages were relatively short, the boats were not undertaking their normal work, and their performance could only be assessed on one particular day, invariably in the winter season. Thus the representativeness of our observations needs to be confirmed by further fieldwork. Moreover, we had to work through interpreters who were, initially at least, unfamiliar with boatbuilding and seafaring terms. With a good interpreter this need not be a handicap, indeed in some respects it can be an advantage: interpreters who knew little about boats and the sea have proved to be among the best. Nevertheless, there must inevitably be some loss of information, if not entire misunderstanding, when questions and answers pass through a third party.

Future research

Future investigators may profit from our experiences. A truly comprehensive documentation of a traditional boat type would have to take a much longer-term view. The research plan would have to ensure that the design process, and the thinking behind it, could be studied directly. All phases of the building process, from choosing the raw material to launching and fitting out, would similarly have to be directly observed and documented in detail. Our experiments with a video camera showed that it was invaluable in resolving certain problems when writing up the research back in Europe, but it could not replace the traditional methods: measured drawings, technical notes and photography.

Other research which needs to be undertaken over a period of time includes: studying the tasks the boats are used for throughout the seasons of the year; and evaluating the boat's performance in different roles and in different wind and sea conditions. Answers would be sought to such questions as: how close to the wind; how much leeway; what speeds; what loads; and why was this sailing rig, and that method of steering, used on a particular hull? Ample time would have to be allocated to finding, recording and understanding boats with features that varied from those found to be the norm in a particular tradition.

For such a research programme to be successful, the investigators would have to learn the regional language, and possibly a local coastal dialect, and spend up to a year on fieldwork on the boats of one tradition. Such a lengthy enquiry can present insoluble problems for Europeans. However, near-comprehensive documentation might be achieved, perhaps in a piecemeal fashion, by involving a number of experienced South Asian scholars, each undertaking fieldwork in their own linguistic region.

It is possible that the group of Indian archaeologists and historians who took part in the basic training course held at Tamil University in January 2000 (McGrail, 2001B) could form the basis for such a research group. They have been trained in the elements of documenting a traditional boat, and some of them are now intent on gaining further experience by recording boats in their own states. They should also be able to encourage others to become involved in this research and, as two or three of them are in senior positions, they may also be able to draw official attention to the urgency of this task.

If such an enthusiastic group of practical scholars (and similar groups in the other countries of South Asia) were to receive official support, they could, in due course, complement the work already undertaken by the Boats of South Asia project. In addition to the near-continuous monitoring of the design and building process, observing boats at work throughout the year and noting variations from procedures and techniques already documented, they could seek further information about a number of specific topics which have not been fully discussed in our published accounts of the three traditions. Among these are: the lofting floor method of designing the *thoni* (especially urgent as the technique appears to be rapidly being replaced by the use of standardised moulds); the factors determining the time of day when fishing boats put to sea and the time that they return; and the distribution of hide boats in the twenty-first century.

Other research fields which need to be investigated include:

A   The documentation of log rafts throughout their zone of use.
B   Area surveys so that types of water transport may be selected for detailed documentation:
    (a)   along the Bangladesh coast to the border with Burma (Myanmar), a boundary which also appears to mark a distinct change in style of boatbuilding;
    (b)   the Sri Lankan coast and the western coast of India and Pakistan;
    (c)   the headwaters of some of the great river systems – for example, the Krishna-Godavari in Andhra Pradesh.

A fundamental enquiry

Mention has been made in Chapter 1 of the value of boat ethnographic research in South Asia to studies in European archaeology and history (see also the Appendix to this volume). There is another field of enquiry which could be similarly rewarding to scholars in both continents and, indeed, would be of value worldwide. During the five years of the Boats of South Asia project we have attempted, in a non-rigorous way, to answer a fundamental question: why are certain types of raft and boat used from one landing place, whereas other types are used from another landing place with (at least superficially) similar characteristics? And, conversely, why are similar types of raft and boat used from landing places which appear to have significantly different characteristics? Such questions have a universal application, both in time and space: answers would be of great value when interpreting excavated boats and their contexts.

To attempt to answer such questions would first entail identifying a number of variables which are thought to influence the selection of specific types of water transport for specific landing places and then defining a number of states for each variable. Examples of such variables, under the broad heading 'Location', are: geomorphology, predominant wind seasonally, beach declivity, raw materials available, types of fish caught etc. The broad heading 'Water transport' could include such variables as: shape, structure, propulsion, function, number in crew etc. While 'People' would include: relative income; religion; social status; tools and techniques; subsidies available. Experience might suggest that many of these variables should be subdivided. A provisional list of variables, with their states, would have to be tested on a pilot survey, then refined before formal fieldwork began.

A small-scale proto-survey in southern Tamil Nadu during January 2000 has underlined the complexity of such an investigation. For example: for any results to have more than local relevance, the distorting effects of government subsidies, and of maverick fishermen and boatbuilders, would have to be discounted in some way. Further consultations are clearly necessary with specialists in disciplines such as anthropology, meteorology and coastal geography. More theoretical studies are also required before a first draft of the variables and states can be compiled. If this can be achieved, it is thought that field trials in southern Tamil Nadu should provide data which, after analysis, could form the basis for a larger survey.

Archaeological excavation in South Asia?

Our present, somewhat sketchy, knowledge of South Asia's maritime past would undoubtedly be enlarged by archaeological surveys of selected maritime regions in South Asia, followed by excavation of specific sites chosen for their deduced importance. Surveys by non-intrusive means might well prove rewarding, providing that the necessary instruments, and their operators and interpreters, were available. Excavation, on the other hand, cannot be contemplated until large-scale conservation facilities for water-logged wood and other 'wet' finds are readily available, along with trained personnel and the long-term resources needed for such scientific centres. Furthermore, the control of the environmental conditions in which such objects have to be displayed is more difficult, and hence more expensive, in South Asia than in more temperate climates. But such facilities need to be, at least, firmly in the planning pipeline before any maritime excavation is mooted.

The recording of traditional South Asian boats by South Asians, as proposed above, should in time result in a body of professionals experienced in dealing with water transport made from organic materials, able to document such rafts and boats in difficult circumstances, and familiar to a degree with the arts and crafts of fishermen and boatbuilders. Such people would have many of the attributes required in a maritime archaeologist. If, after the establishment of the necessary infrastructure, maritime excavations were to be undertaken in the future, archaeologists, historians and others trained as maritime ethnographers would have an important part to play.

APPENDIX: REVERSE-CLINKER PLANKING AND HULC PLANKING PATTERNS, WORLDWIDE

Reverse-clinker planking

Reverse-clinker planking is so called to distinguish it from European clinker, a term which has had long usage. Boats are said to have reverse-clinker planking when each succeeding strake overlaps inboard the upper edge of the strake below, as in the boats discussed in Chapters 2 and 3. In European clinker, on the other hand, the upper plank overlaps *out*board of the lower (see Fig. 2.3 above). Boats with this latter form of planking are also found in India (Greenhill, 1971: 107–109); these are also said to have 'European clinker' planking, but there is no implication in this usage that this is European in origin.

Distribution

Considered worldwide, reverse-clinker planking is very rare, being found nowadays only in South Asia, the USA, Sweden and Britain.

South Asia

Reverse-clinker planking is found in northeastern Bangladesh and in northern Orissa, with a recent 'overspill' into southern West Bengal.

USA

Between *c*. 1955 and 1972, E. MacKenzie built hard-chined, keeled, power boats (sometimes known as 'Cuttyhunk bassboats') at several sites in New England (Purdy, 1997). These boats had European clinker upper planking and reverse-clinker bottom planking which ran from the transom stern to the stem post. They appear to have been built upside down (at least the reverse-clinker parts were), probably on moulds. Purdy noted that the reverse-clinker planking had been incorporated 'on a whim'. An advertisement in *Wooden Boat* (140, Jan/Feb 1998: 130) shows such a boat ('a 26ft MacKenzie') being replanked in a yard at Monument Beach, Massachusetts.

Sweden

Some years ago, before the influx of GRP (glass-reinforced plastic), a few small boats were built with reverse-clinker bottom planking in southern Sweden (J.-O. Traung, pers. comm.).

Britain

Sometime before 1970 (possibly beginning in the 1930s), the See family of Fareham, Hampshire, built two successive classes of racing dinghy with reverse-clinker planking which used to be sailed from the Signal Station at Lee on Solent (McKee, 1972: 23; Leather, 1973: 102–103). These keeled boats were built upside down on a mould and were planked up from sheer to keel; all the planks, except the garboard and the next one, ran from the transom to the sheer (see illustration of a motorised version in *Classic Boat*, 142 [April 2000]: 61).

Advantages of reverse-clinker planking

During fieldwork in Bangladesh (see Chapter 2), builders and crew of reverse-clinker boats were asked why their boats had that sort of planking. Answers were of the form: 'the boat has a better performance', or 'tradition', or 'there is no other way'. Similar replies were received from Orissan and West Bengali boatbuilders and fishermen (see Chapter 3).

In recent times, European and American authors of books on boatbuilding have tried to answer a related question: what are the advantages of reverse-clinker planking over European clinker? Purdy (1997) believed that the use of reverse clinker gave the MacKenzie boats a soft, cushioned ride and increased their directional stability. Traung (pers. comm.), Leather (1973) and McKee (1972) all considered that, in theory, reverse clinker should result in a faster boat. Traung noted that the aim of the Swedish boatbuilders was to introduce air in order to 'lubricate' the boat's bottom, thereby reducing drag. McKee also considered that drag would be decreased, while Leather noted that reverse clinker could allow water to pass outboard instead of being trapped under the lands (laps). Leather also quoted the views of others that there would be an additional lifting force and that there would be increased resistance to leeway (drift downwind). McKee thought that reverse-clinker planking would resist damage when beaching. Both McKee and Leather considered that the most obvious benefit of this form of planking was that it would be easier to clean the inside of the boat, since gunge would not accumulate on the tops of the lands, as it does in a boat with European clinker planking, but would fall into the bilges. McKee further considered that any external caulking would be easier since it could be driven in from above. McKee and others have also pointed out the theoretical disadvantages to reverse-clinker planking: less spray suppression and therefore a wetter boat in a seaway; and less resistance to roll.

It is noteworthy that none of these theories appears to have been investigated by comparative trials or tank tests. Furthermore, none of these four authors appears to have built a boat with reverse-clinker planking nor, as far as is known, sailed in one. On the evidence at present available, it appears that neither planking method is technically superior to the other or results in a boat with a significantly better performance.

It seems likely that, in the past, reverse clinker and European clinker were developed separately within distinct cultural systems. Cultural continuity ensured that once one of the two methods had been adopted, the alternative was not considered: there may never have been a time (until recently) when a choice between the two methods could be made. Theoretically, reverse-clinker methods might tend to arise when building a boat with a plank-keel (moulded dimension less than 0.70 of the siding): the garboard strakes would overlap (i.e. stand on) the inner face of the plank-keel. On the other hand, European clinker might be evolved when building a boat with a keel (moulded dimension greater than 0.70 of siding): a better landing for the garboards could be found on the outer faces of the keel.

Reverse-clinker planking in medieval Europe

There are at present two sites where excavators have thought that reverse-clinker planking may have been exposed.

Utrecht

In 1974 fragments of a vessel known as 'ship II' were excavated on a medieval site at Utrecht (Hoekstra, 1975). These were said to be the 'remains of a small reverse-clinker built ship'. Nearby was another large fragment of ship planking, 'again built in reverse-clinker'. No further evidence has so far been published and the remains do not appear to have been recovered.

Morgan's Lane, Southwark, London

A group of incomplete strakes excavated from a late-sixteenth-century waterfront site in 1987 (Marsden, 1996: 136–44) has been interpreted as being 'from the stern of a boat that had been built in reverse-clinker'. The interpretation of such isolated fragments, not fastened to other timbers, depends on which way the planks are orientated. In the only examples of nailed reverse-clinker boats (ancient or modern) published in detail – those of twentieth-century Orissa (see Chapter 3) – the nails are driven from inboard and clenched outboard (Fig. 2.3 lower). This is good woodworking practice, as the plank being fastened is driven towards the planking already fastened in the hull. With European clinker, it is correspondingly good practice to drive the nails from outboard.

If the Morgan's Lane fragments are orientated with the nail heads inboard, as in reverse-clinker planking, and taking into account the direction of the scarfs in Marsden's planks B and G, it can be deduced that the planking was from the starboard side of its parent vessel, with plank A uppermost. On the other hand, if these fragments are orientated with the nail heads outboard, as for European clinker, the planking was from the port side, with plank H uppermost. Without further information it is not possible to choose between these two possibilities.

In the light of present knowledge of reverse-clinker boats, it seems unlikely that this technique can be infallibly recognised in clinker planking excavated in isolation. However, there are other circumstances when identification may be possible. Finds of overlapping planking fastened to other timbers which can be indisputably orientated can certainly be identified. Isolated finds of framing timbers which can be orientated, and which have been joggled to fit the laps in clinker planking, would also be an indication of European or reverse-clinker planking, depending on the direction of the notches.

Representations of reverse-clinker planking

In India

Reverse-clinker planking is represented on a twelfth-century stone relief in the Jagannath temple at Puri, Orissa. Other stone reliefs depicting this technique, and believed also to be from twelfth-century Orissa, are now in the Victoria and Albert Museum, London (see Fig. 3.3), and in the India Museum, Calcutta (see Fig. 3.4). All three vessels clearly have reverse-clinker planking, but other details of their structure and their method of propulsion are difficult to interpret. The Antwerp artist Solvyns depicted a reverse-clinker *pettoo-a* in the late eighteenth century (see Fig. 3.4) and illustrations by subsequent artists may also depict this technique.

In Europe

Ships evidently with reverse-clinker planking are depicted in several northwest European medieval manuscripts and on town seals from the twelfth to the fifteenth century (see Chapter 3). Among the most convincing of these depictions are the following: a drawing dated before 1140 in *John of Worcester's Chronicle*; a drawing in the *Life of St Thomas of Canterbury* of *c.* 1240; the common seal of Haverford West of 1291; and the seal of the Admiralty Court of Bristol of *c.* 1446. This type of planking is also depicted on a stone frieze in the Chapter House of Salisbury Cathedral (McGrail, 2000).

Hulc planking patterns

In the full hulc pattern, as seen in some representations of medieval European hulcs (Greenhill, 1995A: 250–255; figs 319 and 323; 2000), all lower strakes rise towards both ends of the vessel so that they terminate, not at a post, but at a level or angled surface, well above the waterline. In this configuration, the first strake on each side runs (almost) the full curved length of the vessel. All other strakes of this nature are shorter in length and do not approach the post. Further strakes, running the full length of the vessel, may be fitted above these lower strakes, sealing off, as it were, their ends.

Such a planking pattern is a feature of the Orissan/Bengali *patia* (see Fig. 3.7) and the *Sylheti nauka* (see Figs 2.11, 2.17); it is also a vestigial feature of the *barki* (see Figs 2.13, 2.20) and the *Chhataki nauka* (see Fig. 2.14). The twentieth-century boats with reverse-clinker planking from Britain and the USA, on the other hand, with a conventional stem and a transom stern, do not have the full hulc pattern (the details of the Swedish boats are unknown). The Lee on Solent boats seem to have had hulc planking at the bow since most of their reverse-clinker planks were laid 'from the transom to the sheer'. However, the bottom planking of the American MacKenzie boats runs from the transom to the stem post in a 'conventional' plank pattern.

Medieval European representations of hulc planking

Ships with hulc planking are depicted on a number of town seals, coins, stone carvings, paintings and manuscript illustrations dating from the twelfth to the sixteenth century. These seem to fall into three classes:

(A) *Hulc planking at both ends*. The planking runs to a (near) horizontal surface on which there may be a figurehead or a castle. All strakes end above the waterline.

See Greenhill (2000; 1995A: figs 323, 322, 319; 1995B: figs 1.1, 1.4, 2.1, 9.2, 18.15 to 17), Friel (1995: fig. 2.4) and Hutchinson (1994: frontispiece, fig. 1.5). McKee's hypothetical hulc (Greenhill, 1995A: fig. 54) and Hutchinson's conjectural hulc (1994: fig. 1.7) are also of this type.

(B) *Hulc planking at the bow only*. At the stern the planking appears to run to a conventional post.

See the seal of the Admiralty Court of Bristol (Greenhill, 2000, fig. 2; 1995A: fig. 321), the second seal of Southampton (Friel, 1995: fig. 2.6) and the seal of Thomas Beaufort of *c*. 1416 to 1426 (Greenhill, 2000: fig. 3).

(C) *Hulc planking in upper hull only.* The lower planking runs to hog, post or transom.

See Greenhill (1995A: fig. 320; 1995B: figs 9.1, 32.11), Hutchinson (1994: figs 3.2, 3.6) and Friel (1995: fig. 5.3). Manning's hypothetical hulc (Greenhill, 1988: 64) is of this type.

Of the eighteen and more examples noted above, only five depict nails in a way that suggests reverse-clinker planking.

Recent ships and boats with hulc planking

Netherlands

Some Dutch seagoing craft of the eighteenth and nineteenth centuries seem to have had a complex form of partial hulc planking (Adams *et al.*, 1990: 52–3; Menzel, 1997: 80–81, 90). Ship fragments recovered from sites SL 1, 3 and 5 in the Rhine/Meuse estuary during large-scale dredging in 1986 show that this vessel, built *c.* 1798, was from a Dutch tradition of small merchant traders with round ends, possibly a *tjalk*, a *smak* or a *kuff* (Adams *et al.*, 1990: figs 58, 60 and 61; Menzel, 1997: fig. 91). The ship was built plank first with flush planking held edge to edge by temporary clamps. Subsequently framing was inserted and the planking permanently fastened to it. The lower planking was evidently laid in hulc fashion; this was topped by four or five strakes running the full length of the vessel, from post to post; then more hulc planking; and finally several strakes running near-horizontally from post to post. A model of a *kuff* built in 1869, now in the Prins Hendrik Museum, Rotterdam, has a less complex arrangement with the lowest strakes running between the posts, then four strakes of hulc planking topped by four or five full-length, near-horizontal strakes (Menzel, 1997: fig. 91). A reconstruction drawing of the Dutch *kof* the *Vrouw Maria*, wrecked in Finnish waters in 1771, has hulc planking at the bow (*Nautica Fennica*, Helsinki, 2000: 8–9).

Sweden

A 'falbat fran Malax' built by Bertil Bonns and illustrated in the newsletter of the Swedish Maritime Museum (*Sjohistoriska*, 2/98) appears (possibly at the bow only) to have a planking pattern similar to that of the Dutch *kuff*, but using European clinker, with battens fastened outboard under each lap of the lower hull.

South America

In NE Brazil the lower planking of a keeled boat with a small, raised transom stern, probably built frame first and upside down, has two flush-laid strakes each side running full length from stem to transom; then four hulc strakes. These are capped by the upper planking of four full-length, near-horizontal strakes (observation by Colin Palmer in 1988).

North America

Flat-bottomed cargo vessels of the late nineteenth century, used on the River Pistcataque in New England and known as *gundalow*, had hulc planking at bow and stern (Sam Manning, pers. comm.). The main source is a model in the Smithsonian Institution, Washington, DC: see Howard I. Chapelle's *Catalogue of the National Watercraft Collection*: 103–105, published in 1960 (Basil Greenhill, pers. comm.).

Britain

The Severn trow *Safety*, built in Stourport in 1838, was sketched on the Hubbastone tidal slip at Appledore in 1951 by Vernon Boyle. This drawing, in the possession of Basil Greenhill, shows that this frame-built trow, of 45 register tons, had a dozen or so strakes of hulc planking topped by twelve or more full-length, near-horizontal strakes.

South Asia

The known South Asian reverse-clinker boats with hulc planking patterns are discussed above in Chapters 2 and 3. There are, however, several boat types in South Asia which have a hulc planking pattern but are not reverse clinker (Greenhill, 2000: 12–13, figs 9, 10). In Bangladesh these include the *patam* (Greenhill, 1971: fig. 11) and the *pallar* (see Fig. 2.7 above; see also Greenhill, 1971: figs 9 and 10); in northern Orissa and West Bengal the *salti* and the *chhoat* (Mohapatra, 1983). These all have hulc planking at both ends with ten or so hulc strakes capped by five or so near-horizontal, full-length strakes. Some of the sewn plank boats (*masula*) of Andhra Pradesh also have hulc planking (see Fig. 5.11).

Why hulc planking?

It is clear from this brief and selective review that hulc planking and its variants is (and has been) widely, albeit sparsely, used in South Asia, Europe and the Americas. It is also clear that it is possible to build in reverse-clinker and not necessarily lay the planking hulc fashion. Conversely, hulc planking does not have to be laid in reverse-clinker fashion, though it may be; it can also be used with edge-fastened European clinker and half-lap planking, and with non-edge-fastened flush-laid planking. Moreover, although it seems to be found mostly in plank-first vessels, it can also be used in those built frame first.

To date, no one who has actually built a vessel with hulc planking has explained why that pattern was chosen. Nevertheless it is possible to consider what reasons builders might have had when they chose this planking pattern. The builder of a reverse-clinker vessel with high ends, such as the *patia*, too large to be conveniently built upside down, seems to have little, if any, choice: because of the difficulty of fastening reverse-clinker planking to posts it would seem impracticable to build with any plank pattern other than hulc planking. For the more general case, several possible reasons have been suggested (Leather, 1973: 102–103; Coates, pers. comm.):

a) The problem of fashioning watertight joints between post and underwater strake ends is avoided, since all planking ends above the waterline.
b) The excessively shaped bilge planks of conventional planking patterns are avoided.
c) The planks lie flatter on the vessel's sections and need little twist at the strake ends.
d) The planks are straight and almost of constant breadth for most of their length, an advantage when the breadth of timber available is limited and/or when costs are to be minimised.

A recent small-scale experiment by Basil Greenhill (2000: 3–18) has provided some support for theoretical reason (d) above. A half-model was built to a shape and a planking pattern based on photographs, taken by Greenhill in the 1950s, of a Bangladesh reverse-clinker boat (see Fig. 2.12), with the cross-section of a similar boat measured during recent fieldwork by the Boats of South Asia project (see Fig. 2.21). This model proved to have several strakes which were almost straight. A second model was then built with fuller ends. Strake diagrams based on this model showed clearly that all the planks were 'straight or virtually so'. Greenhill (2000: 15) states that this experiment has shown that a round, full, shallow and capacious hull can be planked hulc style, thus ensuring economy in material and time, since runs of straight, parallel-sided, pit-sawn timber can be used.

BIBLIOGRAPHY

Adams, J., van Holk, A.F.L. and Maarlleveld, Th.J., 1990. *Dredgers and Archaeology.* Rotterdam: Ministerie van Welzijn, Volksgezondheid en Cultuur.

Ahmad, E., 1972. *Coastal Geomorphology of India.* Orient Longman: New Delhi.

Alertz, U., 1995. Naval architecture and oar system of medieval and later galleys. In: Morrison, J. (ed.), *Age of Galley*: 142–62. Conway's History of the Ship, Vol. 2.

Anderson, R.C., 1925. Italian naval architecture c. 1445, *Mariner's Mirror* 11: 135–63.

Arnold, B., 1975. The Gallo-Roman boat from the Bay of Bevaix, Lake Neuchâtel, Switzerland, *International Journal of Nautical Archaeology* 4: 123–41.

Arnold, B., 1978. Les barques celtiques d'Abbeville, de Bevaix and d'Yverdon, *Archeologia* 118: 52–60.

Arulraj, V.S. and Rajamanickam, G.V., 1988. Traditional boats in Tamil literature. In: Rajamanickam, G.V. and Subbarayalu, Y. (eds), *History of Traditional Navigation*: 7–18. Thanjavur: Tamil University.

Arunachalam, B., 1996. Traditional sea and sky wisdom of Indian seamen. In: Ray, H.P. and Salles, J.-F. (eds), *Tradition and Archaeology*: 261–81. Delhi: Manohar.

Barker, R., 1988. Many may peruse us: ribbands, moulds and models in the dockyards, *Revista da Universidade de Coimbra* 34: 539–59.

Barker, R., 1991. Design in the dockyard c. 1600. In: Reinders, R. and Paul, K. (eds), *Carvel Construction Techniques*: 61–9. Oxford: Oxbow Mono 12.

Barker, R., 1993. J.P. Sarsfield's Santa Clara: an addendum, *International Journal of Nautical Archaeology* 22: 161–5.

Barnaby, K.C., 1967. *Basic Naval Architecture.* Hutchinson.

Bay of Bengal Pilot, 1887. London: Admiralty Hydrographic Series.

Beaudouin, F., 1970. *Les bateaux de l'Adour.* Bayonne.

Begley, V., 1983. Arikamedu reconsidered, *American Journal of Archaeology* 87: 461–81.

Begley, V., 1996. Ancient Port of Arikamedu, *Memoirs Arch.* 22. Pondicherry: Ecole Française d'Extrême-Orient.

Behera, K.S., 1994. Maritime contacts of Orissa: literary and archaeological evidence, *Utkal Historical Research Journal* 5: 55–70.

Bellabarba, S., 1993. Ancient method of designing hulls, *Mariner's Mirror* 79: 274–92.

Bellabarba, S., 1996. Origins of the ancient methods of designing hulls: a hypothesis, *Mariner's Mirror* 82: 259–68.

Bidault, J., 1945. *Pirogues et Pagaies.* Paris: J. Susse.

Bird, E.C.F. and Schwartz, M.L. (eds), 1985. *The World's Coastlines*. New York: Van Nostrand Reinhold.

BIWTA, 1994. *Experimental Project for Improving the Efficiency and Profitability of Country Boat Operation*. Dhaka: Bangladesh Inland Water Transport Authority.

Blake, B.A., 1969. Technological Change among the Coastal Marine Fishermen of Madras State. PhD thesis, University of Wisconsin.

Blake, W.M., 1935. Taking off the lines of a boat, *Mariner's Mirror* 21: 5–13.

Bloesch. P., 1983. Mediterranean whole moulding, *Mariner's Mirror* 69: 305–306.

Blue, L., 2000. An Indian reverse-clinker boatbuilding tradition. In: Litwin, J. (ed.), *Down the River to the Sea*: 183–6. Gdansk: Polish Maritime Museum.

Blue, L., Kentley, E. and McGrail, S., 1998. *Vattai* fishing boats and related frame-first vessels of Tamil Nadu, *South Asian Studies* 14: 41–74.

Blue, L., Kentley, E. and McGrail, S., forthcoming. Ethnography, archaeology and India's maritime past. In: *Proceedings of the Lyon 1996 Conference*.

Blue, L., Kentley, E., McGrail, S. and Mishra, U., 1997. *Patia* fishing boats of Orissa: a case study in ethnoarchaeology, *South Asian Studies* 13: 189–207.

Brodrick, J., 1952. *Saint Francis Xavier*. London: Burns Oates.

Caldwell, R., 1881. *Political and General History of the District of Tinnevelly*. Madras: Government Press.

Carrell, T.L. and Keith, D.H., 1992. Replicating a ship of discovery, *International Journal of Nautical Archaeology* 21: 281–94.

Casson, L., 1989. *The Periplus Maris Erythraei*. Princeton University Press.

Cederlund, C.O., 1994. European origin of small water craft in N. America. In: Westerdahl, C. (ed.), *Crossroads in Ancient Shipbuilding*: 247–8. Oxford: Oxbow Monograph 40.

Chaudhuri, K.N., 1985. *Trade and Civilisation in the Indian Ocean*. CUP.

Chittick, N., 1980. Sewn boats in the western Indian Ocean and a survival in Somalia, *International Journal of Nautical Archaeology* 9: 297–309.

Clarke, D.L., 1978. *Analytical archaeology*. London: Methuen.

Claughton, A., Wellicome, J. and Shenoi, A. (eds), 1998. *Sailing Yacht Design: Theory*. Longman.

Crumlin-Pedersen, O., 1969. *Das Haithabuschiff*. Neumünster.

Crumlin-Pedersen, O., 1972. The Vikings and Hanseatic merchants 900–1450. In: G.F. Bass (ed.), *A History of Seafaring*: 181–204. London.

Crumlin-Pedersen, O., 1982. Comment. In: McGrail, S. (ed.), *Woodworking Techniques before AD 1500*: 73. Oxford: BAR/National Maritime Series 129.

Crumlin-Pedersen, O., 1983. *From Viking Ships to Hanseatic Cog*. Greenwich: National Maritime Museum.

Das, M.N., 1977. *Sidelights on the History and Culture of Orissa*. Cuttack: Vidyapuri.

Davendra, S., 1995. Pre-modern Sri Lankan watercraft: the twin-hulled logboat. In: *Royal Asiatic Society of Sri Lanka Sesquicentennial Commemorative Volume*: 211–38.

De Boe, G., 1978. Roman boats from a small river harbour at Pommeroeul, Belgium. In: J. du Plat Taylor and H. Cleere (eds), *Roman Shipping and Trade: Britain and the Rhine Provinces*: 22–30. London.

De Kerchove, Rene, 1961. *International Maritime Dictionary*. New York: Van Nostrand Reinhold.

Deloche, J., 1983. Geological considerations in the localisation of ancient sea-ports of India, *Indian Economic and Social History Review* 20/4: 439–48.

Deloche, J., 1985. Etudes sur la circulation en Inde. IV. Notes sur les sites de quelques ports anciens du pays Tamoul, *Bulletin de l'Ecole Française d'Extrême-Orient* LXXIV: 141–66.

Deloche, J., 1994. *Transport and Communications in India Prior to Steam Locomotion. Vol. 2. Water Transport.* OUP: New Delhi. [English translation of 1980 French edition published in Paris]

Deloche, J., 1996. Iconographic evidence on the development of boat and ship structures in India. In: Ray and Salles (eds): 199–224

De Weerd, M.D., 1978. Ships of the Roman period at Zwammerdam/Nigrum Pullum, Germania Inferior. In: J. du Plat Taylor and H. Cleere (eds), *Roman Shipping and Trade: Britain and the Rhine Provinces*: 15–21. London.

De Zeyla, E.R.A., 1958. Mechanisation of fishing craft and the use of improved fishing gear, *Bulletin of the Fisheries Research Station Ceylon* 7.

Doran, Edwin, 1981. *Wangka: Austronesian Canoe Origins.* College Station: Texas A&M University Press.

Dotson, J.E., 1994. Treatise on shipbuilding before 1650. In: Unger, R.W. (ed.), *Cogs, Caravels and Galleons*: 160–168. Conway's History of the Ship, Vol. 3.

Doust, D.J., Hayes J.G. and Tsuchiya, T., 1965. A statistical analysis of FAO resistance data for fishing craft. In: *Fishing Boats of the World* 3. London: Fishing News (Books).

Edye, J.W., 1834. Native vessels of India and Ceylon, *Journal of the Royal Asiatic Society* 1: 4–14.

Edye, John W., 1834. *Mr Edye on the Native Vessels of India and Ceylon.* London: J.W. Parker.

Ellmers, D., 1972. *Frühmittelalterliche Handelsschiffahrt in Mittel- und Nordeuropa.* Neumünster.

Ellmers, D., 1978. Shipping on the Rhine during the Roman period: the pictorial evidence. In: J. du Plat Taylor and H. Cleere (eds), *Roman Shipping and Trade: Britain and the Rhine Provinces*: 1–14. London. Ellmers, D., 1984. Punt, barge or pram – is there one tradition or several? In: S. McGrail (ed.), *Aspects of Maritime Archaeology and Ethnography*: 153–72. Greenwich: National Maritime Museum.

Facey, W., 1979. *Oman: A Seafaring Nation.* Muscat: Ministry of Information and Culture.

Farrer, A.P., 1996. Recording a craft's lines, *Mariner's Mirror* 82: 216–22.

Finch, R., 1976. *Sailing Craft of the British Isles.* London.

Folkard, H.C., 1870. *The Sailing Boat: A Treatise on English and Foreign Boats and Yachts.* London: Longmans Green.

Friel, I., 1995. *Good Ship.* London: British Museum Press.

Fryer, J., 1698. A new account of East India and Persia. In: Wheeler, J.T. (ed.), *Earl's Records of British India* (1861): 54.

Garrett, R. 1996. *The Symmetry of Sailing.* Sheridan House.

Gerritsma, J., Keuning, J.A. and Versluis, A., 1993. Sailing yacht performance in calm water and in waves. In: *Proceedings of The Eleventh Chesapeake Sailing Yacht Symposium.* The Society of Naval Architects and Marine Engineers.

Greenhill, B., 1957. Boats of East Pakistan, *Mariner's Mirror* 43: 106–134, 203–215.

Greenhill, B., 1961. More evidence for the separate evolution of the clinker-built boat in Asia, *Mariner's Mirror* 47: 296–7.

Greenhill, B., 1966. *Boats of East Pakistan.* Reprint of Greenhill (1957) with additional notes. Soc. Nautical Research.

Greenhill, B., 1976. *Archaeology of the Boat.* London: A & C Black.

Greenhill, B., 1988. *Evolution of the Wooden Ship*. London: Batsfords.

Greenhill, B., 1995A. *Archaeology of Boats and Ships*. London: Conway Maritime.

Greenhill, B., 1995B (ed.). *Evolution of the Sailing Ship*. London: Conway.

Greenhill, B., 1971. *Boats and Boatmen of Pakistan*. Newton Abbot: David & Charles.

Greenhill, B., 2000. The mysterious hulc, *Mariner's Mirror* 86: 3–18.

Grenier, R., 1988. Basque whalers in the New World. In: Bass, G.F. (ed.), *Ships and Shipwrecks of the Americas*: 69–84. London: Thames & Hudson.

Grenier, R., Loewen., B. and Proulx, J.-P., 1994. Basque shipbuilding technology c. 1560–1580: the Red Bay project. In: Westerdahl. C. (ed.), *Crossroads in Ancient Shipbuilding*: 137–41. Oxford: Oxbow Monograph 40.

Gulbrandsen, Ø., 1979. *Beachcraft Development, Andhra Pradesh, India*. Madras: Development of Small-Scale Fisheries in the Bay of Bengal RAS/040/SWE, Project report IND/BCD/1.

Gulbrandsen, Ø., Gowing, G.P. and Ravikumar, R., 1980. *Technical Trials of Beachcraft Prototypes in India*. Madras: Development of Small-Scale Fisheries in the Bay of Bengal BOBP/WP/7 GCP/RAS/040/SWE.

Gurtner, P., 1960. Development of a boat for India's surf coasts. In: J.-O. Traung (ed.), *Fishing Boats of the World* 2: 585–95. Farnham: Fishing News Books.

Guy, J., 1995. Sculptural boat model from eastern India. In: Allchin, R. and Allchin, B. (eds), *South Asian Archaeology, 1995*, Vol. 2: 769–78.

Hall, B., 1868. *Lieutenant and Commander*. London: Bell & Daldy.

Hardgrave, R.L., 2001. *Boats of Bengal*. New Delhi: Manohar.

Hasler, H.G., 1950. Balance-board boats of Palk Strait, *Mariner's Mirror* 36: 265–6.

Hawkins, C.W., 1965. Tuticorin Thoni, *Mariner's Mirror* 51: 147–55.

Hawkins, C.W., 1977. *Dhow*. Lymington: Nautical Publishing.

Hill, A.H., 1958. Some early accounts of the Oriental boat, *Mariner's Mirror* 44: 201–217.

Hoekstra, T.J., 1975. Utrecht, *International Journal of Nautical Archaeology* 4: 390–92.

Hoerner, S.F., 1965. *Fluid – Dynamic Drag*. Hoerner Fluid Dynamics, Vancouver.

Hornell, J., 1910. Report on the feasibility of operating deep sea fishing boats on the coasts of the Madras presidency, with special reference to the selection of fishing centres and harbours of refuge, *Madras Fishery Bulletin* 4: 33–70.

Hornell, J., 1916. Report on pearl fishery held at Thoni, 1914, *Madras Fisheries Dept. Bull.* 8: 43–92.

Hornell, J., 1920. Origins and ethnological significance of Indian boat design, *Memoirs of the Asiatic Society of Bengal* 7/3: 139–256.

Hornell, J., 1922. Indian pearl fisheries of the Gulf of Mannar and Palk Bay, *Madras Fisheries Dept. Bull.* 16: 1–188.

Hornell, James, 1923. Survivals of the use of oculi in modern boats, *Journal of the Royal Anthropological Institute* 53: 289–321.

Hornell, J., 1924A. Boats of the Ganges, *Memoirs of the Asiatic Society of Bengal* 8/3: 173–98.

Hornell, J., 1924B. Fishing methods of the Ganges, *Memoirs of the Asiatic Society of Bengal* 8/3: 201–238.

Hornell, James, 1924C. *The Fishing Methods of the Madras Presidency – Part 1, the Coromandel Coast*. Madras: Government Press.

Hornell, J., 1926. Edye's account of Indian and Ceylon vessels in 1833, *Mariner's Mirror* 12: 45–68.

Hornell, J., 1930. Tongue and groove seam of Gujarati boatbuilders, *Mariner's Mirror* 16: 310–12.

Hornell, J., 1933. Coracles of south India, *Man* 33: 157–60.

Hornell, J., 1943. Fishing and coastal craft of Ceylon, *Mariner's Mirror* 29: 40–53.

Hornell, J., 1944. The outrigger canoes of Madagascar, East African and the Comoro Islands, *Mariner's Mirror* 27: 54–68.

Hornell, J., 1945. Pearling fleets of south India and Ceylon, *Mariner's Mirror* 31: 214–30.

Hornell, James, 1946. *Water Transport: Origins and Early Evolution*. Cambridge: CUP. Second impression, 1970. Newton Abbott: David & Charles.

Horridge, G.A., 1979. *Western Ship in an Eastern Setting*. Greenwich: Maritime Monograph 39.

Howe, J.D.G.F. and Gifford, E.W.H., 1981. *Study of the Operational and Technical Efficiency of Inland Country Boats in Bangladesh*. Government of People's Republic of Bangladesh, Dhaka.

Hunter, W.W., 1872. *Orissa*, Vol. I. London: Smith Elder.

Hutchinson, G., 1991. Early 16th century wreck at Studland Bay, Dorset. In: Reinders, R. and Paul, K. (eds), *Carvel Construction Techniques*: 171–5. Oxford: Oxbow Monograph 12.

Hutchinson, G., 1994. *Medieval Ships and Shipping*. London: Leicester University Press.

Hydrographer of the Navy, 1978. *Bay of Bengal Pilot*. 10th edition. HMSO.

Jansen, E.G. and Bolstad, T., 1992. *Sailing against the Wind*. Dhaka: University Press.

Jansen, E.G., Dolman A.J., Jerve, A.M. and Rahman, N., 1989. *The Country Boats of Bangladesh*. Dhaka: University Press; also London: Intermediate Technology Publications.

Johnston, S., 1994. Making Mathematical Practice. PhD thesis, University of Cambridge.

Kalavathy, V. and Tietze, U., 1984. *Artisanal Marine Fisheries in Orissa: Techno-demographic Study*. Madras: FAO Bay of Bengal Programme BOBP/WP/29.

Kalyanasundaram, V., 1943. Changes in level in the S.E. coast of Madras, *Indian Geographical Journal* 18: 30–36.

Kapitan, G., 1987–9. Records of native craft in Sri Lanka, *International Journal of Nautical Archaeology* 16: 135–48; 17: 223–36; 18: 137–50.

Kapitan, G., 1991. Record of native craft in Sri Lanka: the single outrigger fishing canoe *oruwa*, *International Journal of Nautical Archaeology* 20: 23–32.

Katupotha, J., 1990. Sea level variations: evidence from Sri Lanka and S. India. In: Rajamanickam, V. (ed.): 53–79.

Keith, D.H., 1988. Shipwrecks of the explorers. In: Bass, G.F. (ed.), *Ships and Shipwrecks of the Americas*: 45–68. London: Thames & Hudson.

Kentley, E., 1985. Some aspects of the masula surf boat. In: S. McGrail and E. Kentley (eds), *Sewn Plank Boats*: 303–318. Oxford: BAR Series 276.

Kentley, E., 1993. Sewn boats of the Indian Ocean: a common tradition? In: J. Coles, V. Fenwick and G. Hutchinson (eds), *A Spirit of Enquiry: Essays for Ted Wright*: 68–71. Exeter: Wetland Archaeology Research Project Occasional Paper 7.

Kentley, E., 1996. The sewn boats of India's east coast. In: Ray, H.P. and Salles, J.-F. (eds), *Tradition and Archaeology: Early Maritime Contacts in the Indian Ocean*, 247–60. New Delhi: Manohar.

Kentley, E. and Gunaratne, R., 1987. *The madel Paruwa* – a sewn boat with chine strakes, *International Journal of Nautical Archaeology* 16: 35–48.

Kentley, E., McGrail, S. and Blue, L., 1999. Further notes on *patia* fishing boats in the Bay of Bengal, *South Asian Studies* 15: 151–8.

Kentley, E., McGrail, S., Palmer, C. and Blue, L., 2000. Further notes on the frame-first vessels of Tamil Nadu, *South Asian Studies,* 16: 143–8.

Keuning, J.A. and Sonnenberg, U.B., 1996. Approximation of the calm water resistance on a sailing yacht based on the 'Delft Systematic Yacht Hull Series'. In: *Proceedings of The 14th Chesapeake Sailing Yacht Symposium*. The Society of Naval Architects and Marine Engineers.

Keweloh, H.W., 1985. Recent boats in the Rhineland. In: C.O. Cederlund (ed.), *Postmedieval Boat and Ship Archaeology*: 315–25. Stockholm.

Kinghorn, A.W., 1996. Tuticorin Express, *Mariner's Mirror* 82: 89–91.

Leather, J., 1973. *Clinker Boatbuilding*. London: Adlard Coles.

Lehmann, L.Th., 1978. The flat-bottomed Roman boat from Druten, Netherlands, *International Journal of Nautical Archaeology* 7: 259–67.

Lipke, P. *et al.* (eds), 1993. *Boats: A Manual for their Documentation*. Nashville, Tennessee: American Association for State and Local History.

Litwin, J., 1985. The development of folk boats in Poland from the examples of structures used on the San and Bug rivers. In: C.O. Cederlund (ed.), *Postmedieval Boat and Ship Archaeology*: 327–49. Stockholm.

Loveson, V.J., Rajamanickam, G.V. and Anbarasu, K., 1990. Remote sensing applications in the study of sea level variation along the Tamil Nadu coast, India. In: Rajamanickam, V. (ed.): 179–90.

Loveson, V.J., Rajamanickam, G.V. and Chandrasekar, N., 1990. Environmental impact of micro-deltas and swamps along the coast of Tamil Nadu, India. In: Rajamanickam, V. (ed.): 159–78.

MacKean, H.R., 1980. *Catching a Line*. St Stephen: New Brunswick Heritage Publications.

Mahalik, N.K., 1992. Evolution of Mahanadi delta and its relation to human history, *Utkal Historical Research Journal* 3: 125–9.

Manguin, P.-Y., 1985. Sewn-plank craft of South East Asia. In: S. McGrail and E. Kentley (eds), *Sewn Plank Boats*: 319–43. National Maritime Museum, Greenwich: BAR International Series 276.

Marsden, P., 1996. *Ships of the Port of London*, Vol. 2. English Heritage Archaeological Report 5.

McGrail, S. (ed.), 1982. *Woodworking Techniques before AD 1500*. Oxford: BAR Series 129.

McGrail, S., 1984. Boat ethnography and maritime archaeology, *International Journal of Nautical Archaeology* 13: 2: 149–50.

McGrail, S., 1993. *Medieval Boat and Ship Timbers from Dublin*. Royal Irish Academy.

McGrail, S., 1995A. Maritime research in India, *South Asian Studies* 11: 175–6.

McGrail, S., 1995B. Romano-Celtic boats and ships: characteristic features, *International Journal of Nautical Archaeology* 24: 139–45.

McGrail, S., 1996. The ship: carrier of goods, people and ideas. In: Rice, E.E. (ed.), *The Sea and History*: 67–96. Stroud: Sutton.

McGrail, S., 1997. Early frame-first methods of building wooden boats and ships, *Mariner's Mirror* 83: 76–80.

McGrail, S., 1998. *Ancient Boats in North-West Europe*. Harlow: Longman.

McGrail, S., 2000. Depictions of ships with reverse-clinker planking in Salisbury Cathedral, *Mariner's Mirror* 86: 452–5.

McGrail, S., 2001A. *Boats of the World: From the Stone Age to Medieval Times*. Oxford: OUP.

McGrail, S., 2001B. The 'Boats of South Asia' project, *South Asian Studies* 17: 221–3.

McGrail, S., Blue, L. and Kentley, E., 1999. Reverse-clinker boats of Bangladesh and their planking pattern, *South Asian Studies* 15: 119–50.

McKee, E., 1972. *Clenched Lap or Clinker*. Greenwich: National Maritime Museum.

McKee, Eric, 1983. *Working Boats of Britain: Their Shape and Purpose*. London: Conway Maritime Press.

Menzel, H., 1997. *Smakken; Kuffen; Galioten, drei fast vergessene Schiffstypen des 18. und 19. Jahrhunderts*. Deutsches Schiffahrt Museums, Band 47. Bremerhaven and Hamburg.

Milne, G. *et al.*, 1998. *Nautical Archaeology on the Foreshore*: 51–67. London: Royal Commission for Historical Monuments (England).

Mohapatra, P., 1983. *Traditional Marine Fishing Craft and Gear of Orissa*. Madras: Development of Small-Scale Fisheries in the Bay of Bengal GCP/RAS/040/SWE.

Moreton, W., Fowles, S. and Peers, R., 2000. Note of a demonstration laser scan of a West African dugout, *Mariner's Mirror* 86: 463–7.

Müllner, A., 1892. Ein Schiff im Laibacher Moor, *Argo* 1: 2–10.

Niyogi, D., 1968. Morphology and evolution of the Subarnarekha Delta, India, *Geografisk Tidsskrift* Supplement 67: 230–40.

Oddy, W.A. (ed.), 1975. *Problems in the Conservation of Waterlogged Wood*. Maritime Monographs and Reports 16. Greenwich: National Maritime Museum.

Palmer, C., 1990. Rig and hull performance, *Wooden Boat Magazine* 92: 76–89.

Palmer, C., Blue, L. and McGrail, S., 2001. Hide boats at Hogenakal on the River Kaveri, Tamil Nadu, *South Asian Studies* 17: 213–21.

Parida, A.N., 1994. Ports of ancient Orissa, *Utkal Historical Research Journal* 5: 71–8.

Pâris, F.-E., 1843. *Essai sur la construction navale des peuples extra-européens*, Vol. 2. Paris: Bertrand.

Patnaik, D., 1982. *Festivals of Orissa*. Bhubaneswar: Orissa Sahitya Akademi.

Patra, K., 1988. *Ports in Orissa*. Bhubaneswar: Panchashila.

Pattanayak, A.K. and Patnaik, J.K., 1994. Pattern of settlement of medieval Orissa – An analysis of its trade and contact (12th century – 1568 AD). In: *Archaeology as an Indicator of Trade and Contact*: 1–12. World Archaeology Congress, New Delhi 1994.

Pearson, C., 1987. *Conservation of Marine Archaeological Objects*. London: Butterworth.

Pepper, J.V., 1981. Harriot's Ms. on shipbuilding and rigging (1608–1610). In: Howse, D. (ed.), *500 Years of Nautical Science, 1400–1900*: 24–8. Greenwich: National Maritime Museum.

Pergolis, R. and Pizzarello, U., 1981. *The Boats of Venice*. Venice.

Pethick, J., 1984. *An Introduction to Coastal Geomorphology*. London: Edward Arnold.

Prins, A.H.J., 1965. *Sailing from Lamu: A Study of Maritime Culture in Islamic East Africa*. Assen: Van Gorcum.

Prins, A.H.J., 1982. The mtepe of Lamu, Mombasa and the Zanzibar Sea, *Paideuma* 28: 85–100.

Prins, A.H.J., 1985. *A Handbook of Sewn Boats*. Greenwich: National Maritime Museum.

Purdy, S., 1997. Hell of a sea boat, *Wooden Boat* 138: 80–87.

Rahman, M.A., 1968. *The Country Boats of East Pakistan*. Dhaka: Planning and Research Cell, East Pakistan Inland Water Authority.

Rajamanickam, V. (ed.), 1990. *Sea Level Variation and its Impact on Coastal Environment*. Thanjavur: Tamil University.

Rajamanickam, G.V. and Jayakumar, P., 1991. *Maritime History of South India, Vol. 3. Tamil Nadu.* New Delhi: C.S.I.R.

Rao, S.R., 1970. Shipping in ancient India (from earliest times to 600 AD). In: Chandra, L., Swarup, D., Gupta, S.P. and Goel, S. (eds), *India's Contribution to World Thought and Culture*: 83–107. Madras.

Ray, H.P., 1986. *Monastery and Guild.* Delhi: OUP.

Ray, H.P., 1993. Maritime History of Ancient India: Issues and Reflections. Paper presented at the conference on 'New Directions in Maritime History', Freemantle.

Ray, H.P., 1994. *Winds of Change.* Delhi: OUP.

Ray, H.P. and Salles, J.-F. (eds), 1996. *Tradition and Archaeology: Early Maritime Contacts in the Indian Ocean.* New Delhi: Manohar.

Redknap, M., 1984. *Cattewater Wreck.* Oxford: BAR 131.

Reinders, R., 1979. Medieval ships: recent finds in the Netherlands. In: S. McGrail (ed.), *Medieval Ships and Harbours in Northern Europe.* Greenwich.

Rennell, J., 1793. *Memoir of a Map of Hindustan.* London.

Rieth, E., 1996. *Le Maitre-gabarit, la tablette et le trebuchet.* Paris: C.T.H.S.

Roberts, K.G. and Shackleton, P., 1983. *The Canoe.* Camden, Maine.

Salisbury, W. and Anderson, R.C. (eds), 1958. *Treatise on Shipbuilding and a Treatise on Rigging Written c. 1620–1625.* Soc. Nautical Research Occ. Pub. 6.

Sambasiva Rao, M., Nageswara Rao, K. and Vaidyanadhan, R., 1983. Morphology and evolution of Mahanadi and Brahmani-Baitarani deltas. In: K.R. Dikshit (ed.), *Geomorphology Contributions to Indian Geography*: 261–7. New Delhi: Heritage.

Sarsfield, J., 1984. Mediterranean whole moulding, *Mariner's Mirror* 70: 86–8.

Sarsfield, J., 1985. From the brink of extinction, *Wooden Boat* 66: 84–9.

Sarsfield, J., 1988. Survival of pre-16th century Mediterranean lofting techniques in Bahia, Brazil. In: Filgueras, O.L. (ed.), *Local Boats*: 63–86. Oxford: BAR Series 438.

Schurhammer, G., 1977. *Francis Xavier, His Life, His Times. Vol. 2 (India, 1541–1545).* Rome: Jesuit Historical Institute.

Sclavounos, P.D. and Nakos, D.E., 1993. Seakeeping and added resistance of IACC yachts by a three-dimensional panel method. In: *Proceedings of The Eleventh Chesapeake Sailing Yacht Symposium.* The Society of Naval Architects and Marine Engineers.

Selemke, G., 1973. Die Ausgrabung eines Binnensee Transportschiffes, *Das Logbuch* 9: 21–4.

Severin, T., 1982. *The Sindbad Voyage.* London: Hutchinson.

Solvyns, F.B., 1799. *Les Hindoos.* Paris.

Sopher, D.E., 1965. Indian boat types as a cultural geographic problem, *Bombay Geographical Magazine* 13: 5–19.

Spriggs, J.A., 1994. *A Celebration of Wood.* WARP Occasional Paper 8.

Srinivasan, R. and Srinivasan, V., 1990. Coastal geomorphology of Tamil Nadu – based on Landsat imagery (without field checks). In: Rajamanickam, V. (ed.): 303–312.

Suryanarayana, M., 1977. *Marine Fisherfolk of North-East Andhra Pradesh.* Calcutta: Anthropological Survey of India.

Taylor, D.A., 1988. Contemporary use of whole-moulding in the vicinity of Trinity Bay, Newfoundland. In: Filgueras, O.L. (ed.), *Local Boats*: 87–100. Oxford: BAR Series 438.

Temple, R.C. (ed.), 1905. *A Geographical Account of Countries round the Bay of Bengal, 1669–1679 by Thomas Bowrey.* Hakluyt Society. Reprinted 1997. New Delhi: Munshiram Manoharlal.

Thivakaran, G.A. and Rajamanickam, V., 1994. Indigenous Sailing Traditions of Tamil Nadu. Paper read at a conference in New Delhi.

Toynbee, G., 1873. *History of Orissa*.

Van Doorn, W.G., 1974. *Oceanography and Seamanship*. London: Adlard Coles.

Varadarajan, L., forthcoming. *Boat Typology and Fishing Communities – A Contextual Study in Bengal*. New Delhi: NISTADS.

Verghese, A. 2000. *Archaeology, Art and Religion: new perspectives on Vijayanagara*. New Delhi: Oxford University Press.

Vitharana, V., 1992. *Oru and Yatra*. Colombo: Sri Lanka National Library Service Board.

Wheeler, R.M., Ghosh, A. and Deva, K., 1946. Arikamedu: an Indo-Roman trading station on the east coast of India, *Ancient India* 2: 17–124.

Witsen, Nicolas, 1690. *Architectura Navalis et Regimen Nauticum*. Amsterdam.

Worcester, G.R.G., 1966. *Sail and Sweep in China*. London: HMSO.

Zeiner, P. and Rasmussen, K., 1958. *Report No. 1 to the Government of India on Fishing Boats*. Rome: FAO Report 945 (Project IND/Fi) FAO/58/10/7991.

GLOSSARY

Active frame   any frame which, with others, defines the hull shape – see also Passive frame.

Beam shelf   a longitudinal member fastened inboard of planking or framing to support the ends of crossbeams.

Beating   to sail close-hauled on the wind.

Bevel   a surface which has been angled to make a fit with another.

Bilge   region between the sides and the bottom of a boat – see also Chine.

Bitts   stout, upright timbers in the bows of a boat to which the anchor cable is fastened.

Blind fastening   see Spike.

Bolt rope   rope sewn around the edges of a sail to prevent fraying.

Bottom boards   lengths of planking laid inside the bottom of a boat (often on top of floor timbers) as a flooring.

Bowline   line led towards the bows from the weather (leading) edge of a square sail, when close-hauled, to keep that edge taut and steady.

Bowse   to haul down on a rope.

Butt   a joint in which two members meet edge to edge or, in the case of planks within strakes, end to end.

Caulking   material laid between two structural members to make the junction watertight.

Chine   The transverse shape of the planking where the bottom of a boat meets the sides – see also Bilge.

Chine strake   strake at the turn of the bilge – see also Transition strake.

Cleat   a projecting wooden fitting to which a rope can be made fast.

Clench   to deform or turn the tip of a fastening so that it will not draw out.

Clew   the lower after corner of a fore-and-aft sail.

Close-hauled   to trim sails so that a vessel can sail as close to the wind as practicable.

Crook   a curved limb of a tree which has grown into a shape useful for boatbuilding.

Crossbeam   a structural member extending across a boat.

Displacement the weight of water displaced by a floating hull; this equals the weight of the hull.

Dolly a metal billet held against the head of a boatnail while it is being clenched.

Double-ended said of a boat which is (nearly) symmetrical about the midship transverse plane.

Draught vertical distance between the waterline and the lowest point of the hull.

Fay to fit closely.

Flare said of a boat when her transverse section increases in breadth towards the sheer; the opposite of tumblehome.

Floor timber/floor a framing timber; a transverse structural member, often a crook, extending from turn of bilge to turn of bilge.

Flush-laid planking in which adjoining strakes are butted edge to edge with no visible overlap.

Fore-and-aft sail a sail which, in normal use, is set near the fore-and-aft line of the boat – see also Square sail.

Foundation plank central bottom plank in a flat-bottomed boat.

Frame a transverse structural member, next to the planking, made up of more than one timber.

Frame-first/skeleton-built a form of boatbuilding in which the framework of keel, posts and (some) frames is set up and fastened before the planking is fashioned.

Framing those members which form the boat's framework.

Freeboard height of sides above the waterline; usually measured amidships.

Futtock an element of a frame, fastened to a floor to extend it (near) vertically towards the sheer.

Garboard lowest strake in a keeled boat.

Grommet strands of rope laid up in the form of a ring.

Halyard line to hoist and lower yard and sail.

Hard a foreshore, the surface of which has been consolidated by wooden hurdles, gravel or stone to facilitate boat operations.

Heel (1) the lower end of a mast;
(2) to incline over, temporarily, to port or starboard.

Joggle to cut a notch in a member so that it will fit closely against another.

Keel the lowest and principal longitudinal strength member; usually joined to stems/posts forward and aft – see also Plank-keel.

Kevel head upper end of a framing timber which protrudes above the sheer, and to which ropes may be made fast.

Knee a crook used as a bracket between two structural members set at about right angles to each other.

Lateen sail a fore-and-aft, triangular-shaped sail, bent to a long yard hoisted obliquely to the mast.

301

Leeway   displacement downwind (to leeward) of course steered.

Limber holes   holes cut in structural members which cross the bottom of a boat to allow free passage of bilge water to a position where it may be baled out.

List   an inclination to one side; a permanent heel.

Lug sail   a four-sided, fore-and-aft sail bent to a yard hoisted so that $\frac{1}{4}$ to $\frac{1}{3}$ of the sail area is before the mast.

Mast step   fitting used to locate the heel of a mast.

Master frame   in frame-first building, that frame, usually set (near) amidships, from which the shapes of other frames are derived.

Moulded   the dimension of a post, keel, frame or other framing member measured at right angles to the run of the hull planking – see also Sided.

Naval architectural ratios and concepts – see Annex I in Chapter 9.

Passive frame   any frame which is shaped to match the curvature of the hull planking at its particular station – see also Active frame.

Pay   to cover seams or a boat's hull with tar or other waterproofing substances.

Plank   a component of a strake that is not all in one piece.

Plank-first/shell-built   a form of boatbuilding in which the planking is (partly) erected and fastened before the frames are inserted – see also Frame-first.

Plank-keel   a keel with the ratio of its moulded to its sided dimension less than 0.71 (McGrail, 1998: 112–113).

Rabbet   a groove or channel worked in a structural member to accept another, without a lip being formed.

Reaching   to sail with the wind approximately at right angles to the boat's fore-and-aft line.

Ribbands   long flexible strips of wood running from stem to stern, used during frame-first building to hold the framing timbers in position until replaced by planking; these also enable the builder to visualise hull shape and to fair the framing.

Rocker   fore-and-aft curvature of the bottom of a boat.

Running   to sail with the wind from the sector astern.

Scantlings   dimensions of a timber used in boatbuilding – see Moulded, Sided.

Scarf   a tapered or wedge-shaped joint between two timbers of similar section at the join.

Seam   junction of two structural members required to be watertight.

Settee sail   a form of lateen sail with a short leading edge (luff).

Sheer/sheerline   longitudinal curve of the upper edge of the hull.

Sheer strake   uppermost strake in a boat.

Sheet   line used to trim the foot of a sail.

Shore   stout timber used to support a vessel during building or after she has taken the ground (beached).

Short grain   formed when the axis of a worked timber is not parallel to the grain (the run of the main fibres).

Shroud   rope leading from the masthead to the sides of the boat to support the mast athwartships.

Sided   the dimension of a post, keel, frame or other framing timber measured parallel to the run of the hull planking – see also Moulded.

Spike/blind nail   a fastening which does not penetrate right through the structural members being joined.

Spile   to transfer a curved line from the hull of a boat being built on to a board, and thus mark out the shape of a frame or a plank.

Square sail   a sail which, in normal use, is set at near right angles ('square') to the fore-and-aft line of the boat. Usually rectangular in shape.

Squat   attitude of a boat when her stern settles lower in the water (due to increased speed), thereby causing increased resistance to forward motion.

Stations   notional reference positions, regularly spaced along the length of a boat.

Stay   rope leading forward or aft from the masthead to support the mast.

Stocks   temporary wooden supports on which boats are usually built.

Strake   a single plank or a combination of planks which stretches from one end of the boat to the other.

Stringer   longitudinal member of a boat's structure, fastened inboard of planking or framing; may be used as a beam shelf.

Tack: (1) to alter course through the wind, at intervals, so that the wind is alternately on the port or starboard bow – see also Wear;
   (2) the lower forward corner of a fore-and-aft sail.

Take against   when one timber is contiguous with another, but not fastened or joined to it.

Timber   used generally to refer to any piece of wood used in boatbuilding; in the plural, sometimes used as a synonym for framing.

Transition strake   strake at the transition between bottom and sides of a boat – see also Chine strake.

Treenail   wooden peg or through fastening used to join two members.

Tumblehome   said of a boat when her transverse section decreases in breadth towards the sheer; the opposite to flare.

Waney-edged   said of a timber or plank converted from its parent log with some sapwood.

Wear   to alter course by bringing the stern through the wind so that the wind is now on the other side of the vessel – see also Tack.

Yard   a spar, suspended upon a mast, and from which a sail is set.

Yard arm   the extreme ends of a yard.

GENERAL INDEX

Illustrations are indicated by *italics*. References to common structural components are indexed only when a special feature is involved.

thwarts 84, 229
see also crossbeams
timber
see wood
time taken to build 126, 246
tools, boatbuilding 56, 57, 113, 148, 171, 174, 214
see also building aids
tourism
effects on *parical* 253, 256
towing 60, 62, 92
trade
with Mediterranean 69, 187, 189
with South East Asia 69, 189
see also Arikamedu; cargo boats
traditional wooden boats
see country boats
transition strake
see chine strake
trawlers 129, 228
typical boat, choice of
Bangladesh 18–19, 39–40, 47–48
hide boat 246, 247
masula 136
Orissa 73
Tamil Nadu 198, 201, 258

underwater excavation
see maritime excavations

Victoria & Albert Museum, London 64, 71
Von Portheim Stiftung, Heidelburg 257

wadding
see caulking
waterproofing 48, 50, 57, 78, 84, 162, 217, 247
weather
see climate
wood
degradation of 71
imported from Kerala 178, 179
preservation of 45, 63, 71
species used in Bangladesh 48, 110
species used in *masula* 126, 137, 145, 146, 151, 152
species used in Orissa 70, 78
species used in Sri Lanka 171
species used in Tamil Nadu 194, 211, 212, 213, 220
supply of 48, 70, 106

Xavier, F. 189, 190

yard 89, 90
see also mast; rigging

GEOGRAPHICAL INDEX

Maps and other illustrations are indicated by *italics*.

INDEX OF SOUTH ASIAN SHIP, BOAT AND RAFT TYPES